Digital Signal Processing

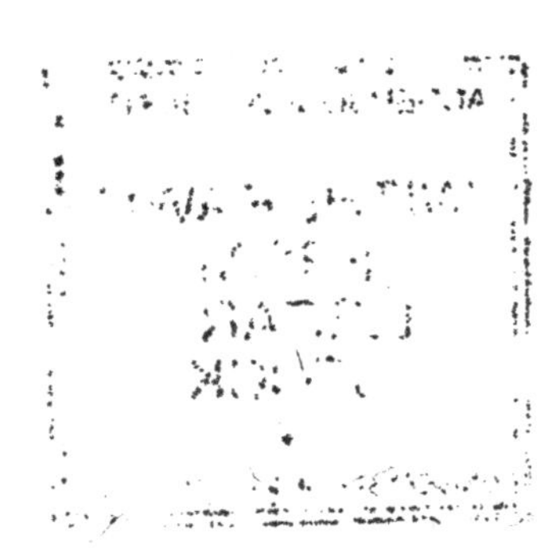

Selected Papers from the
International Conference on Digital Signal Processing,
Facoltà di Ingegneria, Università di Firenze, Italy
August 30 to September 2, 1978

Digital Signal Processing

Edited by

V. Cappellini and **A. G. Constantinides**

Istituto di Elettronica, Facoltà di Ingegneria, Università degli Studi, Florence, Italy
Department of Electrical Engineering, Imperial College of Science and Technology, London, England

1980

ACADEMIC PRESS

A Subsidiary of Harcourt Brace Jovanovich, Publishers

London New York Toronto Sydney San Francisco

ACADEMIC PRESS INC. (LONDON) LTD
24/28 Oval Road,
London NW1

United States Edition published by
ACADEMIC PRESS INC.
111 Fifth Avenue,
New York, New York 10003

British Library Cataloguing in Publication Data

International Conference on Digital Signal Processing, *University of Florence, 1978*
Digital signal processing.
1. Signal processing—Digital techniques—Congresses
I. Title II. Cappellini, Vito III. Constantinides, Antony George
621.3815 TK5101.A1

ISBN 0-12-159080-1 LCCCN 79-41560

Typeset in Malta by Interprint Limited
Printed in Great Britain

List of Contributors

D. Aboutajdine, *Laboratoire d'Electronique et d'Etude des Systèmes Automatiques, Faculté des Sciences, Rabat, Morocco.*

B. P. Agrawal, *ITT Telecommunications Technology Center, 1351 Washington Boulevard, Stamford, Connecticut 06902, U.S.A.*

Z. E. A. Amri, *Laboratoire d'Electronique et d'Etude des Systèmes Automatique, Faculté des Sciences, Rabat, Morocco.*

L. Baroncelli, *Istituto di Elettronica, Facoltà di Ingegneria, Università degli Studi di Firenze; Via di S. Marta 3, 50139 Firenze, Italy.*

S. Bellini, *Centro di Studio per le Telecomunicazioni Spaziali del C.N.R., Via Pouzio Milano, Italy.*

G. Benelli, *Istituto di Elettronica, Facoltà di Ingegneria, Università degli Studi di Firenze, Via di S. Marta 3, 50139 Firenze, Italy.*

M. Bernabo, *Istituto di Elettronica, Facoltà di Ingegneria, Università degli Studi di Firenze, Via di S. Marta 3, 50139 Firenze, Italy.*

A Del Bimbo, *Istituto di Elettronica, Facoltà di Ingegneria, Università degli Studi di Firenze, Via di S. Marta 3, 50139 Firenze, Italy.*

F. Bonzanigo, *Institut für Technische Physik, ETH Hönggerberg, DH 8093 Zürich, Switzerland.*

J. P. Brafman, *Programa de Engenharia Electrica, COPPE-UFRJ, Caixa Postal 1191, ZC-00, 20–000 Rio de Janeiro, Brazil.*

R. Bruno, *Scuola Normale di Pisa, Pisa, Italy.*

G. Bucci, *Centro di Studio per l'Interazione Operatore-Calcolatore, Università di Bologna, Bologna, Italy.*

C. Cafforio, *Istituto di Elettrotecnica ed Elettronica, Politechnico di Milano, Via Pouzio Milano, Italy.*

V. Cappellini, *Istituto di Elettronica, Facoltà di Ingegneria, Università degli Studi di Firenze, Via di S. Marta 3, 50139 Firenze, Italy.*

L. Cristiani Testi, *Istituto di Storia dell'Arte, Università degli Studi, Pisa, Italy.*

A. G. Constantinides, *Department of Electrical Engineering, Imperial College, Exhibition Road, London SW7 2BT, England.*

R. C. Creasy, *European Space Research and Technology Center, Noordwijk, Domeinweg, The Netherlands.*

E. DEPRETTERE, *Network Theory Section, Department of Electrical Engineering, Delft University of Technology, Delft, The Netherlands.*

P. DEWILDE, *Network Theory Section, Department of Electrical Engineering, Delft University of Technology, Delft, The Netherlands.*

E. DUBOIS, *INRS-Telecommunications, University of Quebec, Verdun, Canada.*

P. EDHOLM, *Department of Electrical Engineering, Linköping University, 58183 Linköping, Sweden.*

V. FERRAZZUOLO, *SMA S.p.A., Via del Ferrone, Firenze, Italy.*

M. FONDELLI, *Istituto di Ingegneria Civile, Facoltà di Ingegneria, Università degli Studi di Firenze, Via S. Marta 3, 50139 Firenze, Italy.*

G. V. FRISK, *Woods Hole Oceanographic Institution, Woods Hole, Massachusetts 02543, U.S.A.*

R. A. GABEL, *University of Colorado, Boulder, Colorado 80302, U.S.A.*

G. GARIBOTTO, *Centro Studi e Laboratori Telecomunicazioni, Via G. R. Romoli* 274, 10148 *Torino, Italy.*

B. GOLD, *MIT, Lincoln Laboratory, Lexington, Massachusetts 02173, U.S.A.*

G. GRANLUND, *Department of Electrical Engineering, Linköping University, 58183 Linköping, Sweden.*

R. M. GRAY, *Department of Electrical Engineering, Stanford University, Stanford, California 94305, U.S.A.*

S. HORVATH JR., *Institut für Technische Physik, ETH Hönggerberg, CH 8093 Zürich, Switzerland.*

L. P. JAROSLAVSKI, *Institute for Infömation Transmission Problems, Ermolovoy str, 19 Moscow 103051 U.S.S.R.*

H. KNUTSSON, *Department of Electrical Engineering, Linköping University, 58183 Linköping, Sweden.*

D. MAIO, *Centro di Studio per l'Interazione Operatore-Calcolatore, Universita di Bologna, Bologna, Italy.*

D. R. MARTINEZ, *MIT-WHOI Joint Program in Oceanography/Oceanographic Engineering, Woods Hole, Massachusetts 02543, U.S.A.*

R. M. MERSEREAU, *School of Electrical Engineering, Georgia Institute of Technology, Atlanta, Georgia 30332, U.S.A.*

S. K. MITRA, *Department of Electrical and Computer Engineering, University of California, Santa Barbara, California 93106, U.S.A.*

M. NAJIM, *Laboratoire d'Electronique et d'Etude des Systemes Automatique, Faculte des Sciences, Rabat, Morocco.*

Y. NEUVO, *Tampere University of Technology, Pyynikitine 2, 33230 Tampere 23, Finland.*

A. V. OPPENHEIM, *Research Laboratory of Electronics, Department of Electrical Engineering and Computer Science, MIT, Cambridge, Massachusetts 02139, U.S.A.*

D. Pelloni, *Institut für Technische Physik. ETH Hönggerberg, CH 8093 Zürich, Switzerland.*

M. Piacentini, *Centro di Studio per le Telecomunicazioni Spaziali del C.N.R., Via Pouzio, Milano, Italy.*

R. Pieroni, *Istituto di Elettronica, Facoltà di Ingegneria, Università degli Studi di Firenze, Via di S. Marta 3, 50139 Firenze, Italy.*

A. D. Polydoros, *Department of Electrical Engineering, National Technical University of Athens, Odos 28 Octovriou 42, Athens, Greece.*

J. G. Postaire, *Laboratoire d'Electronique et d'Etude des Systèmes Automatique, Faculté des Sciences, Rabat Morocco.*

A. D. Protonotarios, *Department of Electrical Engineering, National Technical University of Athens, Odos 28 Octovriou 42, Athens, Greece.*

T. A. Ramstad, *Department of Electrical Engineering, Norwegian Institute of Technology, N-7034 Trondheim NTH, Norway.*

P. J. W. Rayner, *Department of Engineering, Cambridge University, Cambridge, England.*

E. Del Re, *Istituto di Elettronica, Facoltà di Ingegneria, Università degli Studi di Firenze, Via di S. Marta 3, 50139 Firenze, Italy.*

F. Rocca, *Istituto di Elettrotecnica ed Elettronica, Politechnico de Milano, Via Pouzio Milano, Italy.*

H. Rudin, *Helsinki University of Technology, Helsinki, Finland.*

I. Saltini, *SMA S.p.A., Via del Ferrone Firenze, Italy.*

K. Shenoi, *ITT Telecommunications Technology Center, 1351 Washington Boulevard, Stamford, Connecticut 06902, U.S.A.*

U. Steimel, *Institut für Informatik der Universität Bonn, Kurfürstenstrasse 74, 53 Bonn, German Federal Republic.*

J. Szczupak, *Centro de Pesquisas de Energia Electrica, Cidade Universitária—Ilha do Fundao, Rio de Janeiro Guanabara, Brazil.*

S. Tazaki, *Department of Electronic Engineering, Ehime University, Matsuyama 790, Japan.*

E. Tzanettis, *Department of Electrical Engineering, Imperial College, Exhibition Road, London SW7 2BZ England.*

A. N. Venetsanopoulos, *Department of Electrical Engineering, University of Toronto, Toronto, Ontario M5S 1A4, Canada.*

Y. Yamada, *Department of Electronic Engineering, Ehime University, Matsuyama 790, Japan.*

T. Zanobini Leoni, *Istituto di Storia dell'Arte, Università degli Studi, Pisa, Italy.*

G. Zappa, *Istituto di Elettronica, Facoltà di Ingegneria, Università degli Studi di Firenze, Via di S. Marta 3, 50139 Firenze, Italy.*

Conference Notes

International Conference on Digital Signal Processing
Facoltà di Ingegneria, Università di Firenze, Italy
August 30–September 2, 1978

Sponsors

Imperial College, London, England
Facoltà di Ingegneria, University of Florence, Florence, Italy
Consiglio Nazionale delle Ricerche (C.N.R.), Italy
Istituto di Ricerca sulle Onde Elettromagnetiche del C.N.R. di Firenze, Italy
A.N.I.P.L.A., Sezione di Firenze dell'A.N.I.P.L.A., Italy
A.E.I., Sezione di Firenze dell'A.E.I., Italy
I.E.E.E. Central-South Italy Section
A.I.C.A., Italy
Università Internazionale dell'Arte, Florence, Italy
A.I.T.A., Italy
In cooperation with I.E.E.E. Audio Speech and Signal Processing Society

Conference Co-chairmen

V. Cappellini, Istituto di Elettronica, Facoltà di Ingegneria, University of Florence, Florence, Italy
A. G. Constantinides, Department of Electrical Engineering, Imperial College, London, England.

Steering Committee

F. Carassa (Italy)
N. Carrara (Italy)
E. C. Cherry (England)
E. De Castro (Italy)
G. Francini (Italy)
E. Gatti (Italy)
G. B. Stracca (Italy)
G. Toraldo di Francia (Italy)

International Technical Committee

M. Bellanger (France)
R. E. Bogner (Australia)
R. Boite (Belgium)
F. Bonzanigo (Switzerland)
C. S. Burrus (U.S.A.)
V. Cappellini (Italy)
A. G. Constantinides (England)
A. C. Davies (England)
A. Fettweis (West Germany)
J. Flanagan (U.S.A.)
B. Gold (U.S.A.)
D. J. Goodman (U.S.A.)
T. S. Huang (U.S.A.)
Y. Kamp (Belgium)
M. Kunt (Switzerland)
L. Jaroslavski (U.S.S.R.)
D. Lebedev (U.S.S.R.)
S. K. Mitra (U.S.A.)
R. Okkes (The Netherlands)
A. V. Oppenheim (U.S.A.)
T. W. Parks (U.S.A.)
P. Rayner (England)
F. Rocca (Italy)
H. W. Schüssler (West Germany)
J. C. Simon (France)
M. N. S. Swamy (Canada)
A. Venetsanopoulos (Canada)
K. Zigangirov (U.S.S.R.)

The Conference followed similar ones in a series held in London in 1971 and Florence in 1972 and 1975 on the same topic. The co-chairmen co-ordinated the programme and Session organization in cooperation with the Steering Committee and the International Technical Committee. The Conference was officially opened on August 30, 1978, by Professor E. Ferroni, Rector of the Florence University, and Dr G. Saccenti, President of the Sezione di Firenze dell'A.N.I.P.L.A. Speakers from twenty-two nations presented more than ninety scientific and technical lectures grouped in twenty-three sessions, and covering the following main topics: design methods and techniques, quantization effects, accuracy and stability; multi-dimensional filtering methods, digital image processing; hardware implementation (including new devices such as special high integration digital circuits and CCDs); applications to signal and image processing with particular reference to speech, radar signals, biomedical signals and images, remote sensing and earth resource data, seismic signals and astronomical observations, air traffic control and machine-tool analysis, fine arts and archaeological prospecting.

On the morning of September 1, a special Session was held at the International University of Art in Florence in cooperation with Professor C. L. Ragghianti, who chaired the Session on the application of digital processing methods and techniques to art works.

In the afternoon of September 2 an interesting Round Table was held on "The Impact of Digital Signal Processing in Technical and Industrial Areas", chaired by Professor F. Carassa of the Politecnico di Milano. Several speakers discussed important areas including such technological innovations as high integration chips and microprocessors, speech processing, voice synthesizers, image processing, industrial process control and applications in new areas like mechanics and chemistry.

The Conference was successful from the point of view of the quality and level of the contributions, and also from the large number of participants.

The aim of these Proceedings is to present a collection of papers, contributed at the Conference so as to have a permanent record and also to satisfy the rather high demand for the limited number of copies of presented papers which were produced.

The original idea was to include all contributed papers but owing to restrictions on the length of the resulting book we have been forced rigorously to edit and select so as to embrace as much as possible of the subject without repetition. That a Conference paper is not included here is no reflection on its quality.

We would like to express our thanks to our respective academic institutions for their continuous support and encouragement. In addition, the invaluable help and assistance of the Consiglio Nationale delle Ricerche, IBM Italia, and S.M.A. (Florence) is gratefully acknowleged.

V. Cappellini and A. G. Constantinides

Preface

During the past decade digital methods and techniques for signal processing have increased their importance to the extent that they now not only replace the classical analogue techniques in many relevant areas, but also find applications in many other new areas. This evolution is the result of several factors, including:

(i) The availability of efficient and relatively simple design methods.

(ii) Realistically obtainable high efficiency, permitting better signal processing and analysis.

(iii) Great flexibility in applications; a single digital technique or system being capable of solving problems in many different fields.

(iv) Extremely rapid and impressive technological advances in large-scale and very-large-scale integration (LSI) circuits for multipliers, accumulators and memories, with an increase in their maximum working frequency and a reduction in cost; the introduction of new devices, working in discrete form, such as charge-coupled devices (CCD) and surface-acoustic-wave (SAW) devices.

(v) Advances in computer hardware and software, in particular the introduction of microcomputers and microprocessors and fast array processors, along with a reduction in cost.

These digital techniques gained importance not only in one-dimensional (1-D) signal processing, but also in two-dimensional (2-D) signal processing (digital image processing).

In the areas outlined above digital techniques for filtering sampled signals or images (digital filtering techniques) are currently the most typical and important cases. As a consequence, digital signal processing techniques have been introduced in the last few years and are now rapidly expanding in different and extremely important fields such as speech processing, communications, radar-sonar, aerospace systems, biomedicine, remote sensing, the processing of results of physical experiments, etc.

The papers contained in these Proceedings (a significant part of those presented at the 1978 Conference on Digital Signal Processing) give advanced and novel contributions in these topics and fields of application; some are of synthetic review nature whilst others contain completely new applications. They are conveniently grouped in five parts.

In Part 1 new and advanced methods and techniques for designing 1-D and 2-D digital filters are presented. Finite impulse response (FIR) and infinite impulse response (IIR) digital filters are described with some criteria for spectral factorization. Generalized filters for stochastic filtering and prediction and synthesis of linear transformations enabling design of a variety of new digital filters and wave digital filters are also presented.

In Part 2, 1-D and 2-D transformations, mainly based on Fast Fourier Transforms (FFT), are described, with special and novel properties emerging. In particular a radix-3 FFT with no multiplications, Shifted Discrete Fourier Transforms, fast computation of Toeplitz forms with narrowband conditions, evaluation of circularly symmetric 2-D Fourier Transforms and FFTs for hexagonally sampled data are described.

In Part 3 some interesting aspects of implementation are examined. A very promising implementation approach using finite arithmetic structures is reviewed, while the evaluation of 2-D quantization makes a significant contribution from the point of view of information theory. Very useful considerations for the implementation of digital filters using microprocessors are also presented.

Part 4 is reserved for the expanding area of applications. Indeed four groups of very interesting contributions are presented. Part 4.1 contains applications to speech processing (variable rate speech processing) and to digital communication systems (PCM coding, digital interpolation of stochastic signals, branch filtering, PCM/FDM conversion). In Part 4.2 are presented applications to biomedicine including 2-D Kalman filtering techniques, aspects of 3-D reconstruction and a new computerized emission tomographic system. Part 4.3 covers applications to radar systems and presents advanced software methods of importance sampling in analysis of radar systems, special chirp filtering techniques and interrogation reply scheduling techniques for a discrete address beacon system in air traffic control. Part 4.4 presents other interesting applications of digital signal processing to determining arrival time in explosion seismology.

Finally Part 5 deals with a completely new field, that of art processing. Here we mean not the "computer art" already known, that is the production of art works by means of the computer, but rather the application of novel, sophisticated techniques to important works of art as an objective analysis, to retrieve a reconstruction of the formative processes adopted by the artist. Important and useful results for the preservation of cultural artefacts are emerging, such as the possibility of storing complete art works for measuring their deterioration in time, electronic re-building on a colour display of some destroyed parts for actual colour replacement, and understanding the construction process by depth reconstruction of a specific painting by means of 2-D to 3-D transformations.

The impact of digital signal processing techniques (such as special digital filters) to archaeological research and prospecting are also described with practical examples.

V. Cappellini and A. G. Constantinides

Contents

List of Contributors v
Conference Notes ix
Preface xiii

Part 1: Design Methods and Techniques for 1-D and 2-D Digital Filters 1

On the Design of High-order Linear Phase FIR Filters
D. Pelloni and F. Bonzanigo 3

Iterative Spectral Factorization Algorithms for Digital Filters
R. A. Gabel, Y. Neuvo and H. Rudin 11

A New Adaptive Recursive LMS Filter
S. Horvath Jr. 21

Some Properties of Frequency Dependent Linear Transformations in Digital Filter Design
A. G. Constantinides and E. Tzanettis 27

Generalized Orthogonal Filters for Stochastic Prediction and Modelling
E. Deprettere and P. Dewilde 35

Two-dimensional Phase Filtering
G. Garibotto 47

Part 2: Transformations 47

Applications of Algebraic Numbers to Computation of Convolutions and DFTs with Few Multiplications
A. N. Venetsanopoulos and E. Dubois 61

Shifted Discrete Fourier Transforms
L. P. Jaroslavski 69

Fast Computation of Toeplitz Forms under Narrow Band Conditions with Applications to Spectral Estimation
U. Steimel 75

A Technique for the Evaluation of Circularly Symmetric Two-dimensional Fourier Transforms and its Application to the Measurement of Ocean Bottom Reflection Coefficients
A. V. Oppenheim, G. V. Frisk and D. R. Martinez 87

A Two-dimensional Fast Fourier Transform for Hexagonally Sampled Data
R. M. MERSEREAU 97

Part 3: Implementations 103

The Application of Finite Arithmetic Structures to the Design of Digital Processing Systems
P. J. W. RAYNER 105

Evaluation of Two-dimensional Quantization
S. TAZAKI, Y. YAMADA and R. M. GRAY 117

Topological Considerations in the Implementation of a Digital Filter using Microprocessors
J. P. BRAFMAN, J. SZCZUPAK and S. K. MITRA . . . 123

Part 4: Applications 135

4.1. Applications to speech processing and communications

Variable Rate Speech Processing
B. GOLD 137

Selection of a PCM Coder for Digital Switching
K. SHENOI and B. P. AGRAWAL 147

Digital Interpolation of Stochastic Signals
A. D. POLYDOROS and E. N. PROTONOTARIOS . . . 163

Branch Filtering using FIR and IIR Complementary Structures
T. A. RAMSTAD 171

4.2. Applications to biomedicine

Two-dimensional Kalman Filtering with Applications to the Restoration of Scintigraphic Images
L. BARONCELLI, A. DEL BIMBO and G. ZAPPA . . . 183

Aspects of 3-D Reconstruction by Fourier Techniques
H. KNUTSSON, P. EDHOLM and G. GRANLUND . . . 197

Design of a Computerized Emission Tomographic System
S. BELLINI, C. CAFFORIO, M. PIACENTINI and F. ROCCA . 207

4.3. Applications to radar systems

Some Importance Sampling Techniques with Applications to Radar Systems
G. BENELLI, V. CAPPELLINI, E. DEL RE. 217
V. FERRAZZUOLO, R. PIERONI and I. SALTINI

Digital Chirp Filtering using the Chinese Remainder Theorem
R. C. CREASEY 227

Performance Evaluation of An Interrogation – Reply Scheduling Technique for a Discrete Address Beacon System
G. BUCCI and D. MAIO 237

4.4. Other applications

Arrival Time Determination in Explosion Seismology
D. ABOUTAJDINE, Z. EL ABIDINE AMRI, M. NAJIM and J.-G. POSTAIRE 249

Part 5: Art Processing 257

Edge Extraction Techniques for Analysis of Art Works
M. BERNABÒ, V. CAPPELLINI and M. FONDELLI . . . 259

Analysis and Reconstruction of Art Works using a Digital Computer
R. BRUNO, L. CRISTIANI TESTI and T. ZANOBINI LEONI 269

Application of Digital Image Processing to Archaeological Prospecting
V. CAPPELLINI and M. FONDELLI 277

Subject Index 285

Part 1

DESIGN METHODS AND TECHNIQUES FOR 1-D AND 2-D DIGITAL FILTERS

On the Design of High-order Linear Phase FIR Filters

DANIELE PELLONI and FEDERICO BONZANIGO

Institut für Technische Physik, Hönggerberg, Zürich, Switzerland

1. Introduction

For some applications with very strict specifications high-order linear phase FIR filters have to be designed. Such filters will be either implemented directly as they are, using for example fast convolution techniques[1], or used as prototypes for decimation or interpolation filters[2] or polyphase networks.[3,4]

Nowadays a linear phase FIR filter will be designed by the minimax approximation of the desired amplitude response using some computer program based on the Remez algorithm. Unfortunately the CPU time grows quite rapidly with the filter length N. Typically it will grow with N^2 using the popular program by McClellan[5] for filters with the same specifications. This will make the design of a high-order filter a long and expensive task. As an example, the design of a low-pass filter with $N=512$ takes already about 15 CPU minutes on a CDC 6500. Moreover one will usually not get the definitive design at the first pass: more passes are needed where one has to vary the filter length and the weights until the specifications on passband ripple and stopband attenuation are met. It is therefore desirable to find some way to reduce the computer time needed for the design of a filter. One would be the ability to extrapolate the properties of the high-order filter from the ones of an appropriate lower-order filter: in this way the determination of the order and the weights could be done on a low-order design.

The Remez algorithm works on a set of extremal frequencies, i.e. frequencies where the magnitude of the weighted error shows a relative maximum. We will use here the extremal frequencies of the low-order filter to get better starting values for the design of the high-order filter: in this way the number of iterations of the Remez algorithm will be reduced.

This paper presents an attempt in this direction. It is the result of an empirical investigation done using the McClellan program[5] on an interactive computing facility.

2. An outline of the proposed method

The proposed method derives from the observation that in multi-passband-stopband high-order filters one can distinguish between regions located in the middle of broad passbands or stopbands where the extremal frequencies are quite evenly distributed and regions near the transitions where they are not. If the regions with a uniform distribution of the extremal frequencies are cut out in an appropriate manner we observe the passband ripple and the stopband attenuation of the lower-order filter obtained in this way are very near to the ones of the original high-order filter.

The procedure will be explained by the example of a multi-passband-stopband filter shown in Fig. 1. The regions A, C and E contain uniformly distributed extremal frequencies. They are cut out and the remaining regions B and D joined together in order to obtain the lower order filter shown in Fig. 2.

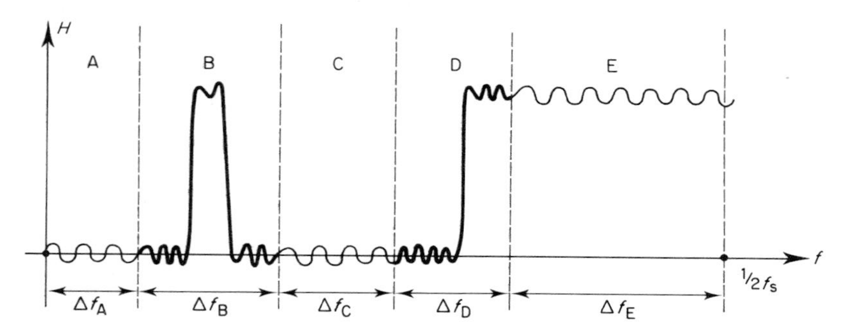

Fig. 1. A high-order filter.

One is now able to design the low-order filter that approximates the specifications shown in Fig. 2. By varying the order and the weights one will find the order and the weights which give a filter showing the desired passband ripple and stopband attenuation. Now the extremal frequencies of the low-order design are plugged back in the corresponding regions on the frequency axis of the high-order filter, the cut-out regions are filled in with evenly spaced extremal frequencies and the whole set of extremal frequencies is used as initial value for a final run of the Remez algorithm that designs the high-order filter.

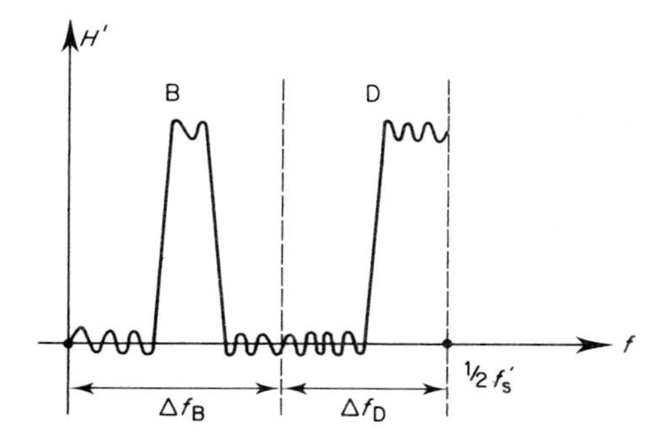

Fig. 2. The lower-order filter obtained from the filter shown in Fig. 1.

3. Relations between the low- and the high-order filter

Now some more details will be given on how the proposed method has to be implemented. In particular some relations between the filter length, the number of extrema and the sampling frequency of the high- and the low-order filters have to be established. We found that it is easier to think in terms of unnormalized frequencies, although it is not mandatory to do so. In the following the variables pertaining to the low-order filter are primed.

First the relationship between the filter orders of the high- and the low-order filter are considered. A relation between the filter length N, the relative transition width ΔF, the passband deviation δ_1 and the stopband deviation δ_2 was given in ref. 7 for low-pass filters and in ref. 8 for bandpass filters:

$$N = \frac{D_\infty(\delta_1, \delta_2)}{\Delta F} - F(\delta_1/\delta_2)\Delta F + 1 \tag{1}$$

where D_∞ and F are functions of the deviations δ_1 and δ_2 and are given in references 7 and 8.

For a very small ΔF, as it is the case with the high-order filters considered here, one gets from equation (1):

$$N \approx \frac{D_\infty(\delta_1, \delta_2)}{\Delta F} \tag{2}$$

If we take the absolute transition width $\Delta f = \Delta F . f_s$ and the deviations δ_1 and δ_2 as constant and make the sampling frequency f_s variable we obtain from equation (2):

$$N \approx \frac{D_\infty(\delta_1, \delta_2)}{\Delta F f_s} f_s = K_N f_s \tag{3}$$

where $K_N = D_\infty / \Delta f$ will be a proportionality constant. We observed that this approximate proportionality holds for multi-passband-stopband filters, too. From (3) we can get the order N' of the low-order filter obtained by cutting out the uniform regions of a high-order filter of order N.

$$N' = N \frac{f'_s}{f_s} \tag{4}$$

For the example of Fig. 1 this amounts to

$$N' = N \frac{\Delta f_B + \Delta f_D}{(\Delta f_A + \Delta f_C + \Delta f_E) + (\Delta f_B + \Delta f_D)}$$

From equation (4) follows that as the cut-out regions are made broader the shorter the low-order filter will result. On the other side, one has to be careful not to cut out regions where the extremal frequencies will no longer be evenly distributed, or else the results will be bad. It is difficult to predict where the extremal frequencies will be still uniform and where they will not. Therefore it is better to be a little too conservative and keep a region where the extrema will turn out to be evenly distributed. Furthermore the parity of the order of both the high- and the low-order filter has obviously to remain the same.

Now we consider the number of extremal frequencies. If we do not consider the particular case of the extraripple filters, the number of extremal frequencies N_e is related to the filter length N by:[9]

$N_e = \frac{1}{2}(N+1)+1$ for odd length, symmetric impulse response

$N_e = \frac{1}{2}N+1$ for even length, symmetric impulse response (5)

$N_e = \frac{1}{2}(N-1)+1$ for odd length, antisymmetric impulse response

$N_e = \frac{1}{2}N+1$ for even length, antisymmetric impulse response

Equations (4) and (5) can be combined. For N even we get

$$(N_e - 1) = K f_s \tag{6}$$

where K is a proportionality factor. For N odd similar expressions would be obtained. However it turned out that it is more convenient to use equation (6) for all cases and to modify equation (4) accordingly.

Now we should see how the extremal frequencies have to be placed as starting values for the high-order filter design. The N'_e extremal frequencies of the low-order design will be moved on the frequency axis with the regions they belong to. As reference points the cutoff frequencies may be taken. In this way some gaps corresponding to the cut-out regions are left out. In these gaps the remaining ΔN_e extremal frequencies will be filled in evenly distributed. From equation (6), it is

$$\Delta N_e = N_e - N'_e = \left(\frac{f_s}{f'_s} - 1\right)(N'_e - 1) \tag{7}$$

Their number in each gap will be proportional to its width. If an end of a gap occurs either at the zero or at the Nyquist frequency, the nearest extremum will be placed at the same distance the zero or the Nyquist frequency as the corresponding extremum of the low-order filter.

4. Examples

We present here a choice of three out of about 50 examples that were run. The specifications and the characteristics of the low-order design, compared with the ones of the resulting high-order design, are presented in Table 1. In Examples 1 and 2 there was an additional constraint: the filter length had to be a power of two. The weights were chosen so that a higher attenuation in the stopband was obtained. It has to be observed that in Example 3 a filter of length 1469 would already scarcely fulfill the given specifications. In Table 2 the iteration count and the CPU time for designing the high-order filter is shown. This was done using the McClellan program[5] on a DEC System 10 equipped with a KL10 central processor.

5. Conclusions

In the previous sections we have shown how a low-order filter has to be chosen so that the latter can be used to estimate the characteristics of a high-order filter. Actually the passband ripple and the stopband attenuation

TABLE 1

	Specifications			Low-order filter				High-order filter	
Example No.		Range (kHz)	Ripple or att. (dB)	Range (kHz)	Order and f_s	Weight	Ripple or att. (dB)	Order and f_s	Ripple or att. (dB)
1. Low-pass	PB	0–1·7	±0·25	0–1·7		1	±0·244		±0·243
	SB	2·3–5·7	60	2·3–5·7	64	32	61·02	2048	61·06
	SB	6·3–256	80	6·3–8	16 kHz	700	87·82	512 kHz	87·86
2. Band-pass	SB	0–143·4	70	0–8·4		500	72·27	2048	72·77
	PB	144·6–147·7	±1	9·6–12·7	128	1	0·998	576 kHz	0·945
	SB	148·3–288	70	12·3–18	36 kHz	500	72·27		72·77
3. High-pass	SB	0–53·0	55	0–3	95	35	55·86	1535	56·56
	PB	53·2–80	±0·5	3·2–5	10 kHz	1	0·476	160 kHz	0·440

TABLE 2

Example No.	Direct iter.	Direct CPU s	Starting low-order iter.	Starting low-order CPU s	Starting final design CPU s
1	14	4757	5	2262	1527
2	13	4766	5	2460	1522
3	14	2550	7	1419	858

of the high-order filter are usually slightly better than estimated from the low-order filter. For the usual specifications (40–100 dB stopband attenuation and 0·1–1 dB passband ripple) the improvement of the stopband attenuation is less than 1 dB for a filter length ratio not exceeding 20.

In section 4 we considered only examples where the specified amplitude response is either 1 or 0 and the weighting function is piecewise constant. However the proposed method is not restricted to such cases. As an example, in ref. 4 it was applied to the design of the transversal part of a recursive polyphase filter. The specifications of the transversal part were similar to the ones of Example 1, however the weighting in the stopband was a nonuniform periodic function. Here all but two of the stopband periods were dropped to obtain the low-order filter. As starting values for the high-order filter design the extremal frequencies of the last "period" were practically repeated on the frequency axis.

We have also shown that the use of the extremal frequencies obtained from the low-order design as starting values for the high-order design reduces the number of iterations and therefore the computing time. With the McClellan program[5] the CPU time for the high-order design is usually reduced by a factor of from 1·8 to 3·3. However there are cases where the reduction is only by a factor of about 1·2 to 2. In these cases the assumed distribution of the extremal frequencies turned out not to be good and one extremum was shifted from one band to another. This occurs very seldom when only one uniform region is cut out, as in Example 1, or two regions with similar specifications are cut out, as in Example 2. It happens quite often indeed when two or more regions with different specifications are cut out as in Example 3.

The reduction of the number of iterations of the Remez algorithm is always more important than the reduction in computing time. It turned out that the CPU time for the last iteration in the McClellan program was longer by a factor of from 3 to even 10 than the other iterations. This is due to the maximum search algorithm used there. The length of the computation of the filter coefficients is also in the same order of magnitude as a Remez iteration because a simple inverse DFT is used. Another problem in the use of the above mentioned program[5] for designing large FIR filters is the memory space used by the three large arrays used for the storage of the desired amplitude and the weighting on a dense grid and the frequency of the grid points themselves. There is therefore room for a further improvement using a program, such as the one mentioned in ref. 6, that does not show these mentioned drawbacks.

Note added in proof

The program mentioned in ref. 6 has been modified and is now able automatically to design high-order FIR filters using the method described in this paper. Filters up to order 4096 have been approximated. A paper about the results is in preparation.

Acknowledgments

The authors wish to thank Prof. E. Baumann for his encouragement and support. Most of the computation was executed on the DEC System 10 of the Center of Interactive Computing (ZIR) of the Swiss Federal Institute of Technology.

References

1. H. D. Helms, "Fast Fourier Transform Method of Computing Difference Equations and Simulating Filters", *IEEE Trans. on Audio and Electroacoustics*, **AU-15**, No. 2, pp. 85–90 (June 1967).
2. R. E. Crochiere and L. R. Rabiner, "Optimum FIR Digital Filter Implementation for Decimation, Interpolation, and Narrow-Band Filtering", *IEEE Trans. on Acoustics, Speech, and Signal Processing*, **ASSP-23**, No. 5, pp. 444–456 (Oct. 1975).
3. M. G. Bellanger, G. Bonnerot and M. Coudreuse, "Digital Filtering by Polyphase Network: Application to Sample-Rate Alteration and Filter Banks", *IEEE Trans. on Acoustics, Speech, and Signal Processing*, **ASSP-24**, No. 2, pp. 109–144 (April 1976).
4. D. Pelloni, "Synthese von digitalen Polyphasennetzwerken", *AGEN Mitteilungen*, No. 24, pp. 1–13 (Dec. 1977).
5. J. H. McClellan, T. W. Parks and L. R. Rabiner, "A Computer Program for Designing Optimum FIR Linear Phase Digital Filters", *IEEE Trans. on Audio and Electroacoustics*, **AU-21**, No. 6, pp. 506–526 (Dec. 1973).
6. F. Braun, A. Kinzl and H. Rothenbühler, "Chebyshev Approximation of Arbitrary Frequency Response for Nonrecursive Digital Filters with Linear Phase", *Electronics Letters*, **9**, No. 21, pp. 507–509 (Oct. 1973).
7. O. Herrman, L. R. Rabiner and D. S. K. Chan, "Practical Design Rules for Optimum FIR Low-Pass Digital Filters", *Bell System Tech. J.*, **56**, No. 6, pp. 769–799 (July-Aug. 1973).

8. F. Mintzer and B. Liu, "An Estimate of the Order of an Optimal FIR Bandpass Filter", *Record 1978 IEEE Int. Conf. on Acoustics, Speech and Signal Processing, Tulsa, OK, April 1978*. pp. 483–486.
9. J. H. McClellan and T. W. Parks, "A Unified Approach to the Design of Optimum FIR Linear-Phase Digital Filters", *IEEE Trans. on Circuit Theory*, **CT-20**, No. 6, pp. 697–701 (Nov. 1973).

Iterative Spectral Factorization Algorithms For Digital Filters

ROBERT A. GABEL

University of Colorado, U.S.A.

YRJÖ NEUVO

Tampere University of Technology, Finland

and

HANS RUDIN†

Helsinki University of Technology, Finland

1. Introduction

In discrete-time signal processing the problem of spectral factorization arises e.g. in the design of filters with prescribed squared-magnitude functions. It is always possible to specify the squared-magnitude function $F(\omega T)$ of a digital FIR-filter‡ with transfer function $H(z)$ by a mirror-image polynomial (MIP) $R(z)$:

$$F(\omega T) = R(z)\Big|_{z=\exp(j\omega T)} = H(z)H(1/z)\Big|_{z=\exp(j\omega T)} \tag{1}$$

In system theory the more general problem of matrix factorization is treated. There are numerous papers which describe approaches to reduce the matrix factorization to the factorization of polynomials.[1] The spectral factorization of polynomials itself, is usually not discussed, even though this part takes most effort from the computational point of view.[2]

These facts provide the motivation to consider here the factorization of a given MIP§:

† On leave from Swiss Federal Institute of Technology

‡ For IIR-filters the same is true for numerator and denominator separately.

§ In principle roots on the unit circle have to be excluded; see however Section 4.

$$R(z) = r_n z^n + \ldots + r_1 z + r_0 + r_1 z^{-1} + \ldots + r_n z^{-n} \quad (2)$$

into its minimum phase factor

$$H(z) = h_0 + h_1 z^{-1} + \ldots + h_n z^{-n} \quad (3)$$

and its maximum phase factor

$$H(1/z) = h_0 + h_1 z + \ldots + h_n z^n \quad (4)$$

2. Approaches to spectral factorization of mirror-image polynomials

Solving a system of nonlinear equations

By multiplying analytically $H(z) = h_0 + h_1 z^{-1} + \ldots + h_n z^{-n}$ with $H(1/z)$ and equating the coefficients of the resulting polynomial with the known coefficients $r_i (i = 0, 1, \ldots n)$ of the given mirror-image polynomial a set of nonlinear equations for the unknown $h_i (i = 0, 1, \ldots n)$ is obtained. Programs for solving nonlinear equations can now be applied. However the solution is not unique and special care must be taken to obtain the minimum phase spectral factor.

Determination of the roots

An obvious method is to calculate all the roots of $R(z)$. Because $R(z)$ is a MIP, these roots occur in reciprocal pairs. All roots inside the unit circle are attributed to the minimum phase spectral factor $H(z)$. The evaluation of all complex roots can be a lengthy process. Numerical sensitivity problems are also encountered in the evaluation of $H(z)$ from its roots. Thus for applications where the roots themselves are not needed, it makes sense to look for methods which directly lead to $H(z)$.

Hilbert transform relationship

Recently Papoulis[3] proposed a spectral factorization method which uses a FFT three times. The method takes advantage of the fact that magnitude and phase of the minimum phase spectral factor form a Hilbert transform pair.

In the following we concentrate on recursive algorithms, which increase the accuracy of the computed spectral factor iteratively.

Bauer's direct iterative method[4]

Bauer performs a LU-decomposition of the infinite Toeplitz matrix $[\boldsymbol{R}]$, which is formed by the given coefficients of the mirror-image polynomial:

$$\begin{bmatrix} r_0 & r_1 & \dots & r_n & & 0 \\ r_1 & r_0 & \dots & & r_n & \\ \vdots & & & & & \ddots \\ r_n & & & & & \\ & r_n & \ddots & & & \\ 0 & & & & & \end{bmatrix} \tag{5}$$

If the resulting triangular matrices are chosen to have identical diagonal elements, the decomposition is unique and the nonzero elements $h_i^{(k)}(i=0, 1, \dots n)$ of the kth row of the upper triangular matrix, or the kth column of the lower triangular matrix respectively, can be calculated recursively:

(1) Set $h_0^{(k)}=\left(r_0-\sum_{m=1}^{n}(h_m^{(k-m)})^2\right)^{1/2}$ (6)

(2) Set $h_i^{(k)}=\left(r_i-\sum_{m=1}^{n-i}h_m^{(k-m)}.h_{i+m}^{(k-m)}\right)/h_0^{(k)}$ for $i=1, 2, \dots n$ (7)

(3) Let $k=k+1$, go to (1)

with initial conditions $h_i^{(m)}=0$ for $m=0, -1, \dots -(n-1)$
With increasing k the elements $h_i^{(k)}(i=0, 1, \dots n)$ converge to the coefficients of the minimum phase factor $H(z)$.

3. The alternate algorithm†

The method described here is based on the well known least-error-energy (LEE) deconvolution filtering—a time-domain approach. It is therefore appropriate to interpret polynomials as time-sequences by taking the inverse z-transform. From $H(z)$ in equation (3) we get the finite length time-sequence $\boldsymbol{h}=\{h_0, h_1, \dots h_n\}$. Correspondingly the coefficients of the given mirror-image polynomial in equation (2) form the autocorrelation sequence $\boldsymbol{r}=\{r_n, r_{n-1}, \dots r_1, r_0, r_1, \dots r_n\}$ of the sequence $\boldsymbol{h}$. This follows from the fact that the multiplication $H(z).H(1/z)=R(z)$ in the z-domain corresponds to the following convolution in the time-domain:

$$r_k=h_k * h_{-k}=\sum_{m=0}^{n}h_m h_{-(k-m)}=\sum_{m=0}^{n}h_m h_{m-k} \quad \text{for } k=0, 1, \dots n \tag{8}$$

which can be recognized as the autocorrelation sequence of $\boldsymbol{h}$.

† Although the algorithm has been used earlier for deconvolution of microseismic records,[5,6] it seems to be unknown in the spectral factorization literature.

In the first step the LEE finite length inverse filter $\boldsymbol{g}$ of the filter with impulse response $\boldsymbol{h}$ is determined. This can be done without knowing $\boldsymbol{h}$, but only its autocorrelation $\boldsymbol{r}$. To the input of the discrete-time system of Fig. 1 a unit sample δ_k is applied. By minimizing the error energy at the output we obtain the following set of equations:

$$\sum_{q=0}^{N} g_q r_{m-q} = \sum_{k=0}^{N} \delta_k h_{k-m} \qquad \text{for } m=0, 1, \ldots N \tag{9}$$

where N is the order of the inverse filter $\mathbf{g}$. Due to the requirement for $H(z)$ to be minimum phase all $h_i \equiv 0$ for $i<0$, so that we can set up the following matrix equation:

$$\begin{bmatrix} r_0 & r_1 \ldots r_n & & & 0 \\ r_1 & r_0 \ldots & r_n & & \\ \vdots & \vdots & & \ddots & \\ r_n & & & & r_n \\ & r_n & \ddots & & \\ 0 & & \cdot r_n \ldots r_1 & & r_0 \end{bmatrix} \begin{bmatrix} g_0 \\ g_1 \\ \vdots \\ \\ g_N \end{bmatrix} = \begin{bmatrix} 1 \\ 0 \\ \vdots \\ \\ 0 \end{bmatrix} h_0 \qquad \text{for } N>n \tag{10}$$

The correlation matrix in equation (10) has Toeplitz structure, so that $\mathbf{g}$ can be evaluated very efficiently by the Levinson algorithm[8]. Levinson's recursion formulae for solving equation (10) are given below.

(1) Compute $v = \sum_{q=0}^{k-1} g_q r_{k-q}$

$$\text{(2) Set} \begin{bmatrix} g_0 \\ g_1 \\ \vdots \\ \\ g_k \end{bmatrix} = \frac{1}{1-v^2} \begin{bmatrix} g_0 \\ g_1 \\ \vdots \\ g_{k-1} \\ 0 \end{bmatrix} - \frac{v}{1-v^2} \begin{bmatrix} 0 \\ g_{k-1} \\ \vdots \\ g_1 \\ g_0 \end{bmatrix} \tag{11}$$

(3) Set $k=k+1$, go to (1) unless $k>N$

with initial conditions $k=1$ and $g_0=1/r_0$

If the LEE inverse filter $\mathbf{g}$ of order N is determined, only one additional recursion step has to be performed to obtain the LEE inverse filter of order $N+1$. From Fig. 1 it is seen immediately that the error energy is zero if $\mathbf{g}$ is the inverse of $\boldsymbol{h}$; i.e. $G(z)=1/H(z)$. With increasing order N, $\mathbf{g}$ approximates the inverse of $\boldsymbol{h}$ more and more accurately.

In the second step the first $n+1$ terms of the division

$$1/(g_0+g_1 z^{-1}+\ldots) = h'_0 + h'_1 z^{-1} + \ldots + h'_n z^{-n} + \varepsilon(z) \tag{12}$$

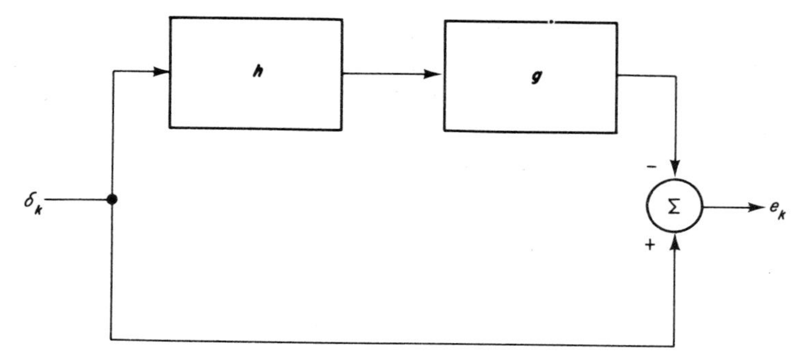

Fig. 1. Least-error energy filtering.

are computed. The resulting coefficients $h'_i(i=0, 1, \ldots n)$ converge (within a scale factor) to the coefficients $h_i(i=0, 1, \ldots n)$ of the minimum phase spectral factor $H(z)$. The final scaling is performed by dividing $h'_i(i=0, 1, \ldots n)$ by $\sqrt{h'_0 h'_n / r_n}$.

EXAMPLE. The following squared magnitude function is given:

$$R(z)=85z^3+622z^2+2491z+8004+2491z^{-1}+622z^{-2}+85z^{-3}$$

Its exact minimum phase spectral factor is equal to

$$H(z)=85+27z^{-1}+7z^{-2}+1z^{-3}$$

By using the algorithm above the first seven iteration steps are performed and the results are listed in Table 1.

4. Convergence, number of numerical operations, sensitivity: a comparison between Bauer's and the alternate algorithm

Speed of convergence

The speed of convergence depends on the distance ρ from the unit circle to the nearest root† of the squared-magnitude function in both algorithms. Bauer states that about $\frac{1}{2}a\log_{10}(1/\rho)$ iteration steps are necessary to approximate the coefficients of $H(z)$ with an accuracy of 10^{-a}. An identical speed of convergence has been found numericaly for the alternate algorithm.

† Theoretically roots on the unit circle have to be excluded from the squared-magnitude functions. In practice an approximate factorization can be obtained with both algorithms. Because the iteratively calculated spectral factor is not exact, its roots are shifted somewhat away from the unit circle. This deviation gets smaller during the iteration process so that the speed of convergence is also decreasing.

TABLE 1.

Step	h_0	h_1	h_2	h_3
1	53·10183042	16·52631929	5·143311013	1·600698118
2	67·55246413	21·46933183	5·391228854	1·258281265
3	84·27643444	27·76969159	6·938476431	1·008585621
4	85·05512774	27·01754319	7·004042273	0·999351859
5	84·99465216	26·99829031	6·999544201	1·000062920
6	85·00000340	27·00000119	6·999999715	0·999999960
7	84·99997804	26·99999302	6·999998042	0·999999960

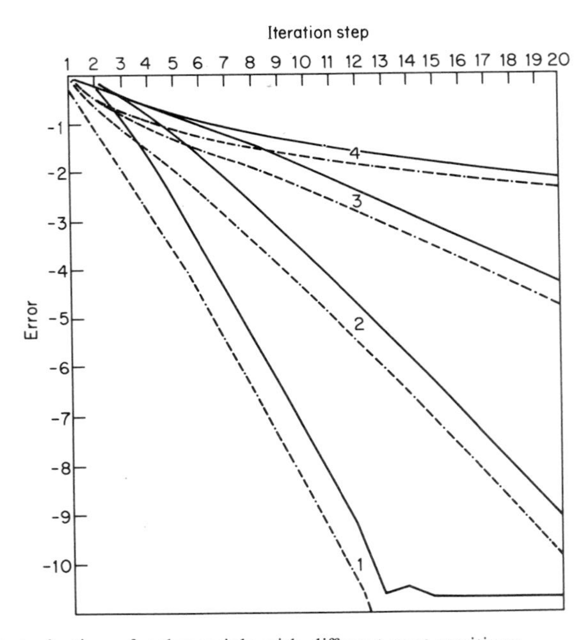

Fig. 2. Factorization of polynomials with different root positions.

$$\text{Error (ordinate)} = \log \frac{\left(\sum_{i=0}^{n} (r_i - r_i')^2\right)^{1/2}}{\sum_{i=0}^{n} (r_i)}$$

Curve 1: $H(z) = (1 - 0{\cdot}3z^{-1})^2$. Curve 2: $H(z) = (1 - 0{\cdot}5z^{-1})^2$.
Curve 3: $H(z) = (1 - 0{\cdot}7z^{-1})^2$. Curve 4: $H(z) = (1 - 0{\cdot}9z^{-1})^2$.
Solid lines indicate the alternate algorithm; dashed lines indicate Bauer's algorithm.

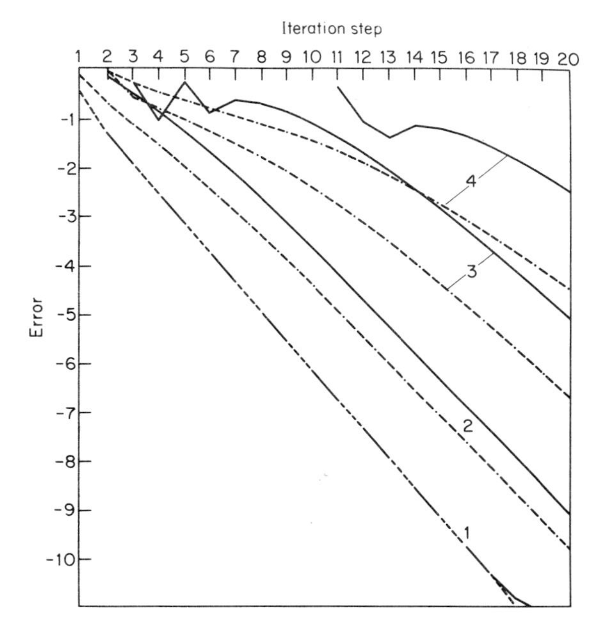

Fig. 3. Factorization of polynomials with increasing order of the critical roots.

$$\text{Error (ordinate)} = \log \frac{\left(\sum_{i=0}^{n} (r_i - r_i')^2\right)^{1/2}}{\sum_{i=0}^{n} (r_i)}$$

Curve 1: $H(z) = (1 - 0{\cdot}5z^{-1})$. Curve 2: $H(z) = (1 - 0{\cdot}5z^{-1})^2$.
Curve 3: $H(z) = (1 - 0{\cdot}5z^{-1})^4$. Curve 4: $H(z) = (1 - 0{\cdot}5z^{-1})^6$.
Solid lines indicate the alternate algorithm; dashed lines indicate Bauer's algorithm.

This is shown in Fig. 2 for fourth order squared-magnitude functions with different ρ-values. The alternate method needs a few more iteration steps to reach the same accuracy as Bauer's algorithm. The speed of convergence remains constant, if the order of the critical roots is increased, but starts on a higher error level. For second order squared-magnitude functions the algorithms produce identical results. This is shown in Fig. 3.

Number of arithmetic operations

For K iteration steps ($K > n$, the order of the spectral factor) the following number of operations have to be performed:

	Bauer's algorithm	*Alternative method*
Multiplications	$(\frac{1}{2}n^2+\frac{3}{2}n)K-\frac{1}{4}(n^2+n+2\sum_{q=1}^{n} q^2)$	$K^2+(n+4)K$ $+(3n+3)$
Additions	$(\frac{1}{2}n^2+\frac{1}{2}n)K-\frac{1}{4}(n^2+n+2\sum_{q=1}^{n} q^2)$	$\frac{1}{2}K^2+(n-\frac{1}{2})K$ $+(n-1)$
Square roots	K	1 (for scaling)

The number of operations increases in Bauer's algorithm linearly with the number of iteration steps K and quadratically with the polynomial order n. The alternative algorithm behaves just vice versa. Figure 4 shows the curves, where the number of operations are equal for both algorithms.

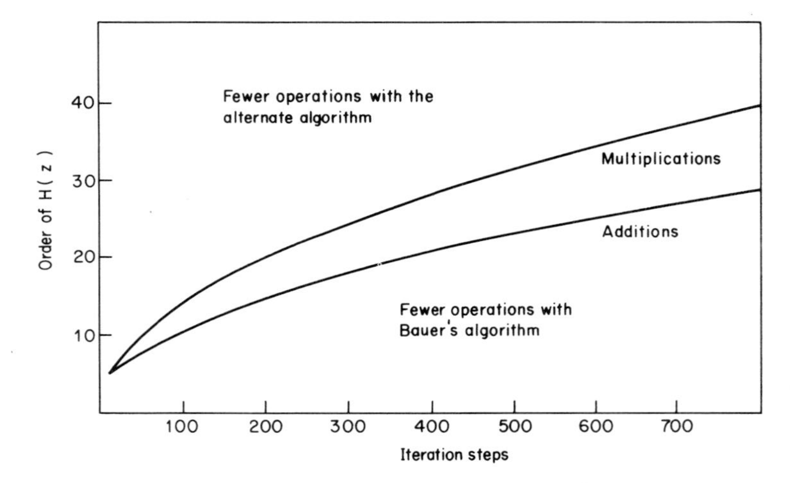

Fig. 4. Comparison of the number of numerical operations.

Sensitivity due to roundoff errors
Bauer's algorithm is self-correcting, which means that convergence is maintained under continued influence of rounding errors. This is not true for the Levinson algorithm. The effect is visible in the curves 1 of Figs. 2 and 3, where the accuracy of the approximations reaches the internal accuracy (13 digits) to three digits of the programmable calculator used (TI-59).

5. Conclusions

Numerical methods for spectral factorization of discrete-time squared-magnitude functions have been discussed. Bauer's method and an algorithm, which is based on least-error energy deconvolution, have been compared. For high-order squared-magnitude functions computational savings can be expected when the alternate algorithm is used.

References

1. B. D. O. Anderson, K. L. Hitz and N. D. Diem, "Recursive Algorithm for Spectral Factorization", *IEEE Trans on Circuits and Systems*, **CAS-21**, No. 6 (Nov. 1974).
2. A. C. Riddle and B. D. Anderson, "Spectral Factorization Computational Aspects", *IEEE Trans. on Automatic Control*, **AC-11** (Oct. 1966).
3. A. Papoulis, "The Factorization Problem for Time-Limited Functions and Trig. Polynomials", *IEEE Trans. on Circuits and Systems*, **CAS-25**, No. 1 (Jan. 1978).
4. F. L. Bauer, "Ein direktes Iterationsverfahren zur Hurwitz-Zerlegung eines Polynomes", *Archiv der elektr. Uebertragung*, **9**, pp. 285–290 (1955).
5. E. A. Robinson, "Statistical Communications and Detection." London. Griffin, 1967.
6. J. N. Galbraith, Computer studies of microseismic statistics with application to prediction and detection, MIT Ph.D. Thesis (1963).
7. N. Levinson, "The Wiener RMS error criterion in filter design and prediction", *in* "Extrapolation, Interpolation, and Smoothing of Stationary Time Series" (ed. N. Weiner) New York, Wiley 1949.

A New Adaptive Recursive LMS Filter

S. HORVATH JR.

Institut für Technische Physik, Hönggerberg, Zürich, Switzerland

1. Introduction

Adaptive recursive filters have been the object of intensive research efforts in recent years.[3-6] The interest in adaptive recursive "least mean-square" (LMS) filters is motivated by the low computational complexity of such filters. For example, Feintuch's adaptive recursive LMS filter[3] has the computational simplicity of adaptive transversal LMS filters. Unfortunately, his LMS algorithm does not in general minimize the mean-square error (MSE), and thus the use of his adaptive recursive LMS filter is greatly limited.[1,2] In this paper, a new LMS algorithm for adaptive recursive filters is proposed that does not have this shortcoming.

2. Derivation of the new recursive LMS algorithm

The adjustable recursive filter structure we consider is described in the time domain by the following input/output relationship:

$$y(n) = \sum_{k=0}^{M} a_k x(n-k) + \sum_{k=1}^{N} b_k y(n-k) \tag{1}$$

where $\{x(n)\}$ is the input sequence, $\{y(n)\}$ the output sequence, and the set of $\{a_k\}$, resp. the set of $\{b_k\}$ are referred to as its feed-forward resp. feedback variable coefficients. The goal is to minimize the MSE between desired output $d(n)$ and actual output $y(n)$ of the filter:

$$E[e^2(n)] = E[(d(n) - y(n))^2] \tag{2}$$

by continuously adjusting the $\{a_k\}$ and $\{b_k\}$ as each new input sample $x(n)$ is received.

Based on the philosophy of steepest descent, the basic iteration in all LMS algorithms is of the form:

$$\boldsymbol{c}(n+1) = \boldsymbol{c}(n) + \mu(-\hat{\boldsymbol{g}}(\boldsymbol{c}(n))) \tag{2}$$

where $\boldsymbol{c}(n+1)$ denotes the new filter coefficient vector, μ is a parameter that controls stability and rate of convergence, and $\hat{\boldsymbol{g}}(\boldsymbol{c}(n))$ is the "noisy" estimate of the gradient of the MSE following the basic idea by Widrow and Hoff.[7] Its ith component is given by:

$$\hat{g}_i(\boldsymbol{c}(n)) = -e(n)\frac{\partial y(n)}{\partial c_i}\bigg|_{\boldsymbol{c}=\boldsymbol{c}(n)} \approx -2E\left[e(n)\frac{\partial y(n)}{\partial c_i}\right]_{\boldsymbol{c}=\boldsymbol{c}(n)} \tag{4}$$

Accordingly, the updating of the recursive filter coefficients is to be carried out as follows:

$$a_k(n+1) = a_k(n) + \alpha e(n)\frac{\partial y(n)}{\partial a_k}\bigg|_{\boldsymbol{a}(n),\,\boldsymbol{b}(n)}, \qquad k=0, \ldots, M \tag{5a}$$

$$b_k(n+1) = b_k(n) + \beta e(n)\frac{\partial y(n)}{\partial b_k}\bigg|_{\boldsymbol{a}(n),\,\boldsymbol{b}(n)}, \qquad k=1, \ldots, N \tag{5b}$$

where $\boldsymbol{a}(n) = (a_0(n), \ldots, a_M(n))$ and $\boldsymbol{b}(n) = (b_1(n), \ldots, b_N(n))$, and α and β are positive constants small enough to insure the convergence of the iterative procedure. The problem is to find a suitably accurate estimate for the required first-order sensitivity functions:

$$\frac{\partial y(n)}{\partial a_k}\bigg|_{\boldsymbol{a}(n),\,\boldsymbol{b}(n)} = x(n-k) + \sum_{i=1}^{N} b_i \frac{\partial y(n-i)}{\partial a_k}\bigg|_{\boldsymbol{a}(n),\,\boldsymbol{b}(n)} \qquad k=0, \ldots, M \tag{6a}$$

$$\frac{\partial y(n)}{\partial b_k}\bigg|_{\boldsymbol{a}(n),\,\boldsymbol{b}(n)} = y(n-k) + \sum_{i=1}^{N} b_i \frac{\partial y(n-i)}{\partial b_k}\bigg|_{\boldsymbol{a}(n),\,\boldsymbol{b}(n)} \qquad k=1, \ldots, N \tag{6b}$$

The difficulty here is that the computation of the actual value of each of them requires, as we see from equation (6), at least NT seconds, with T the time between successive input samples, and constant coefficients, excluding apparently the possibility to perform the coefficient updating sample by sample. In Feintuch's LMS algorithm, the true first-order sensitivity function formulas have been truncated, leaving only the first term of the right-hand side of equation (6), with the result that his algorithm does not in general minimize the MSE.[1] The basic idea is to insure that the new LMS algorithm exhibits a gradient behaviour in the neighbourhood of a stationary point (its behaviour during the transients being of less importance) and to make use of relations that hold in steady state.

It can be shown that in steady state, when the initial conditions can be neglected, the first-order sensitivity functions of the recursive filter considered are related to one another in the following simple manner:

$$\left.\frac{\partial y(n)}{\partial a_j}\right|_{\boldsymbol{a}(n),\,\boldsymbol{b}(n)} = \left.\frac{\partial y(n-j)}{\partial a_0}\right|_{\boldsymbol{a}(n),\,\boldsymbol{b}(n)}, \qquad j=0, \ldots, M \tag{7a}$$

$$\left.\frac{\partial y(n)}{\partial b_j}\right|_{\boldsymbol{a}(n),\,\boldsymbol{b}(n)} = \left.\frac{\partial y(n+1-j)}{\partial b_1}\right|_{\boldsymbol{a}(n),\,\boldsymbol{b}(n)}, \qquad j=1, \ldots, N \tag{7b}$$

Therefore, only the two fundamental sensitivity functions:

$$s_a(n) = \left.\frac{\partial y(n)}{\partial a_0}\right|_{\boldsymbol{a}(n),\,\boldsymbol{b}(n)} \tag{8a}$$

$$s_b(n) = \left.\frac{\partial y(n)}{\partial b_1}\right|_{\boldsymbol{a}(n),\,\boldsymbol{b}(n)} \tag{8b}$$

are needed to determine all first-order sensitivity functions of interest. Furthermore, it can be assumed that near a stationary point the incremental adjustment of the $\{a_k\}$ and $\{b_k\}$ are so small that the filter coefficients are nearly constant:

$$a_k(n) \cong a_k(n-j), \qquad j=1, \ldots, N; \quad k=0, \ldots, M \tag{9a}$$

$$b_k(n) \cong b_k(n-j), \qquad j=1, \ldots, N; \quad k=1, \ldots, N \tag{9b}$$

Based on this assumption, $s_a(n)$ and $s_b(n)$ can be estimated as outlined in equation (6), that we rewrite in the form:

$$\hat{s}_a(n) = x(n) + \sum_{i=1}^{N} b_i(n) \cdot \hat{s}_a(n-i) \tag{10a}$$

$$\hat{s}_b(n) = y(n-1) + \sum_{i=1}^{N} b_i(n) \cdot \hat{s}_b(n-i) \tag{10b}$$

using $\hat{s}_a(n)$ and $\hat{s}_b(n)$ to denote the estimate of $s_a(n)$ and $s_b(n)$, respectively. Note that if the equality sign holds in equation (9), $\hat{s}_a(n)$ and $\hat{s}_b(n)$ assume the true values of $s_a(n)$ and $s_b(n)$, resp., and equation (10) is identical to equation (6). Therefore, by using equation (10) together with equation (7), it is insured that suitably accurate estimates of all the first-order sensitivity functions required in equation (5) are obtained near a stationary point. During the adaptation, equation (9) and therefore equation (10) and the relations described by equation (7) hold only in an approximative manner. However, the estimates become more and more accurate as the stationary point is approached. Such a behaviour has been observed in the applications reported in the next section. Let us finally point out that the relations between the first-order sensitivity functions given by equation (7) are also valid at the start of the adaptation, provided that the recursive filter has zero initial conditions, which is usually the case.

The new LMS coefficient updating algorithm that we suggest is therefore given by:

$$a_k(n+1) = a_k(n) + \alpha e(n)\hat{s}_a(n-k), \qquad k = 0, \ldots, M \tag{11a}$$

$$b_k(n+1) = a_k(n) + \beta e(n)\hat{s}_b(n+1-k), \qquad k = 1, \ldots, N \tag{11b}$$

with $\hat{s}_a(n)$ and $\hat{s}_b(n)$ given by (10*a*) and (10*b*) respectively, and with $\hat{s}_a(n) = \hat{s}_b(n) = 0$ for $n < 0$.

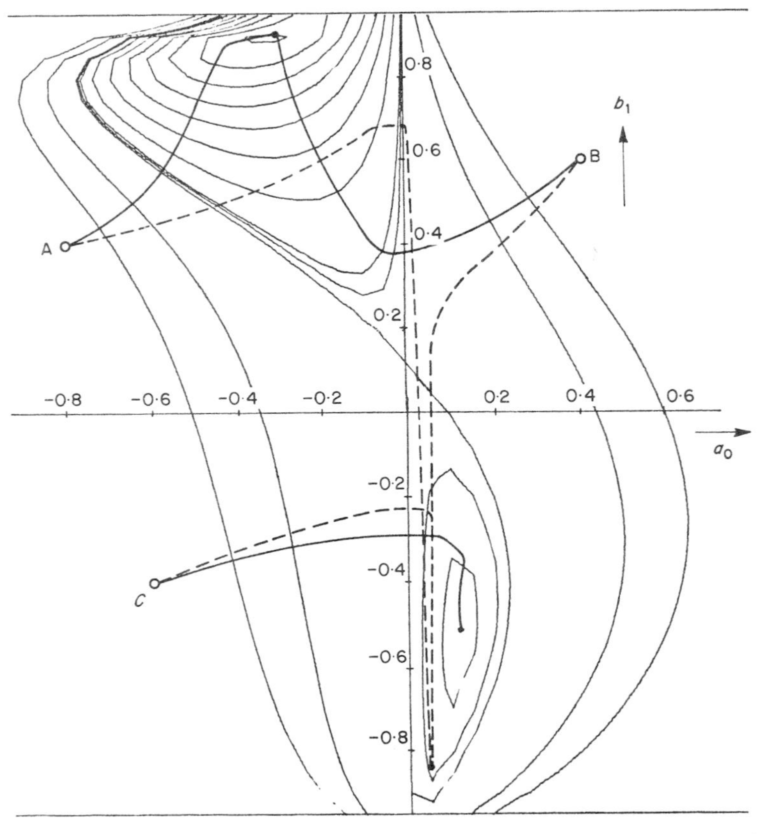

Fig. 1. Bimodal case-contour representation of the MSE (normalized as in ref. 1) showing the locus of the adaptive parameters (a_0, b_1) if the new LMS algorithm is used (continuous line) and if the Feintuch's LMS algorithm is used (dashed line) for three different starting conditions *A*, *B* and *C*. In both algorithms we used $\alpha = \beta = 0{\cdot}00043$. Note that the new algorithm always minimizes the MSE, but can eventually end in a local minimum.

3. Performance of the new adaptive recursive LMS filter and conclusions

The new adaptive recursive LMS filter described by equations (1), (11) and (10) always minimizes the MSE. This is illustrated by comparison to Feintuch's recursive LMS filter as seen in Fig. 1, where the example given by Johnson and Larimore[1] is considered. Figure 2 shows a quantitative performance comparison of the new adaptive recursive LMS filter and the adaptive transversal LMS filter by Widrow,[8] both used as an adaptive time-domain equalizer for high-speed data transmission. The channel to be equalized was a measured telephone channel. Although no proof of convergence is available at the present, these two examples encourage the use of the recursive LMS filter derived in this paper. The convergence properties of the new recursive LMS algorithm are presently the subject of investigations.

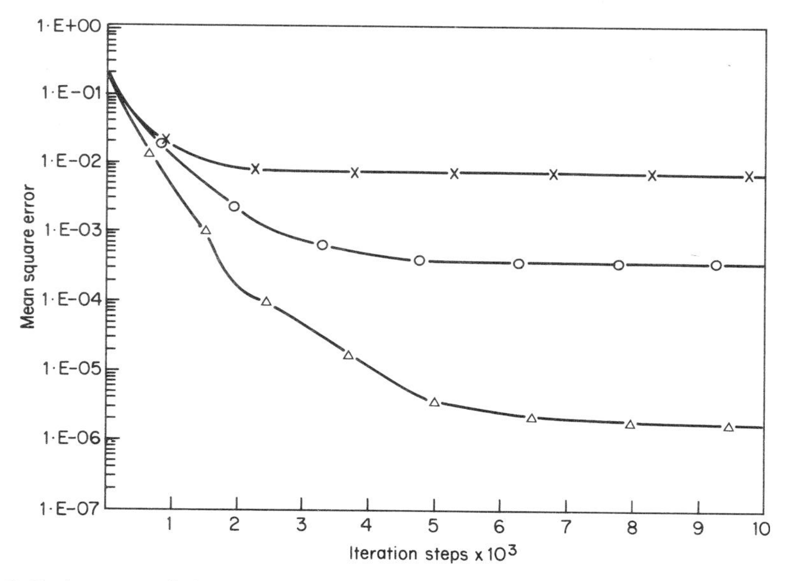

Fig. 2. Performance of the new adaptive recursive LMS filter with $M=N=6$ (curve △) in comparison with that of:

(i) an adaptive transversal LMS filter having the same number (13) of variable coefficients (curve ×),

(ii) an adaptive transversal LMS filter having the same complexity, i.e. requiring the same number of multiplications pro output sample (curve ○). Number of variable coefficients: 19.

All adaptive filters were used as adaptive equalizers. The channel to be equalized was a measured telephone channel.

Acknowledgments

The author is grateful to Prof. Dr. E. Baumann for his constant encouragement and support, to D. Dzung for many interesting discussions and for his help in the computer simulations, and to Dr. R. Lagadec for his help and suggestions in correcting the manuscript.

References

1. C. R. Johnson, Jr. and M. G. Larimore, "Comments on and Additions to 'An Adaptive Recursive LMS Filter'", *Proc. IEEE*, **65**, No. 8, pp. 1399–1402 (Sep. 1977).
2. B. Widrow and J. M. McCool, "Comments on 'An Adaptive Recursive LMS Filter'", *Proc. IEEE*, **65**, No. 8, pp. 1402–1404 (Sep. 1977).
3. P. L. Feintuch, "An Adaptive Recursive LMS Filter", *Proc. IEEE*, **64**, No. 11, pp. 1622–1624 (Nov. 1976).
4. S. A. White, "An Adaptive Recursive Filter", *Proc. 9th Asilomar Conf. on Circuits, Systems and Computers, Pacific Grove, CA, pp.* 21–25, *November* 1975.
5. S. Horvath, Jr., "Adaptive IIR Digital Filters for On-Line Time-Domain Equalization and Linear Prediction", presented at IEEE Arden House Workshop on Digital Signal Processing, Harriman, NY, February 1976.
6. G. R. Elliott, W. L. Jacklin, and S. D. Stearns, "The Adaptive Digital Filter", *Techn. Rept. No. SAND* 76-0360, Sandia Laboratories (Aug. 1976).
7. B. Widrow and M. E. Hoff, Jr., "Adaptive Switching Circuits", in IRE WESCON Conv. Rec., pt. 4, pp. 96–104 (1960).
8. B. Widrow, "Adaptive Filters I: Fundamentals", *Stanford Electron. Lab. Techn. Rept. No.* 6764–4, Stanford University (Dec. 1966).

Some Properties of Frequency Dependent Linear Transformations in Digital Filter Design

A. G. CONSTANTINIDES and E. TZANETTIS

Department of Electrical Engineering, Imperial College, London, England

Introduction

Several methods have been presented in the past for the design of digital filters from doubly terminated ladder filters. The basic work was done by Fettweiss[1,2,3] and later it was extended by Constantinides under a new approach.[4,5] A presentation in a more global form to include all the known methods was first given in ref. 6 but the main emphasis was placed on the applications rather than the theory of the method. Moreover a further generalization in the frequency domain is recently given.[7] The purpose of this paper is to provide a theoretical frame, complete but not detailed, from which the properties of the known methods may be derived and further specific methods may be developed.

1. Frequency dependent linear transformations

The frequency dependent linear transformation applied on the voltage–current variables of a given port i is defined as

$$\begin{bmatrix} X_i \\ Y_i \end{bmatrix} = \begin{bmatrix} a_i & b_i \\ c_i & d_i \end{bmatrix} \begin{bmatrix} V_i \\ I_i \end{bmatrix} = Q_i \begin{bmatrix} V_i \\ I_i \end{bmatrix} \Bigg|_{s = f(z^{-1}),\ |Q_i| \neq 0} \tag{1.1}$$

where a, b, c, d are in general frequency dependent quantities. The X_i, Y_i are the new variables derived from V_i and I_i through equation (1.1). The matrix of transformation Q_i will be referred to as the linear transformation matrix (LTM) for the given port, whose properties in its most general form (frequency dependent) are to be investigated for the different network ports.

Let us consider a doubly terminated network consisting of L cascaded blocks. The properties to be derived result by considering:

(a) each constituent ladder element as a two port network block;

(b) the interconnection between the different two ports;

(c) the terminations of the network;

(d) the derived network as a whole, i.e. its linear stability and the effect of the transformation applied on the s-variable.

2. The linearly transformed two-port

Let us consider the two-port of Fig. 1(a) and its digital equivalent in Fig. 1(b), derived through linear transformations applied on the V and I variables of each port.

Let T be the modified chain matrix (modified in a sense that B and D appear with sign opposite to the normal) and Q_1, Q_2 the LTMs of the ports 1 and 2. In ref. 6 was proved that

$$[X_1 \; Y_1]^T = Q_1 T Q_2^{-1} [X_2 \; Y_2]^T \tag{2.1}$$

We modify equation (2.1) and derive

$$[Y_1 \; Y_2]^T = \sum [X_1 \; X_2]^T \tag{2.2a}$$

where Σ is a 2×2 matrix and

$$\text{DEN} = -afA - bfC + aeB + beD$$

$$\sigma_{11} = \text{DEN}^{-1}(-cfA - dfC + ceB + deD) \qquad \sigma_{12} = \text{DEN}^{-1}(bc - da)(AD - BC) \tag{2.2b}$$

$$\sigma_{21} = \text{DEN}^{-1}(eh - fg) \qquad \sigma_{22} = \text{DEN}^{-1}(-ahA - bhC + agB + bgD)$$

The realization of equations (2.2) gives the appropriate signal flow diagram (SFD). Obviously a realizability condition is that $\text{DEN} \neq 0$. We consider this constraint for two possible kinds of two-ports.

2a. Series impedance, i.e. $A = 1$, $B = -Z$, $C = 0$ and $D = -1$.

By definition we have $|Q_i| \neq 0 (i = 1, 2)$. Since $\text{DEN} \neq 0$ we conclude that a and e may not be both zero.

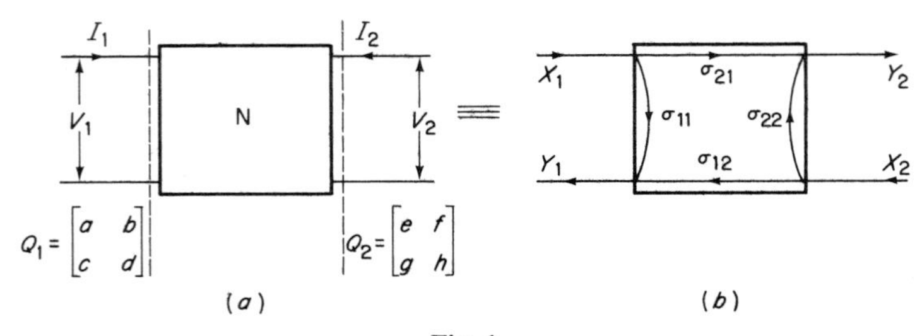

Fig. 1.

2b. Shunt admittance, i.e. $A=1$, $B=0$, $C=Y$ and $D=-1$.
After a similar consideration the same condition is derived for b and f.

3. Interconnection of two-ports

It is achieved through an interconnecting network having mod. chain matrix with $A=1$, $B=0$, $C=0$ and $D=-1$.[6] Given that $|Q_i|\neq 0$ $(i\equiv 1, 2)$ and DEN $\neq 0$ the conclusion is that none of the pairs (a, e) and (b, f) may have both its elements of zero value. The case of simple interconnection should be mentioned here i.e. when

$$X_1=Y_2 \qquad \text{and} \qquad X_2==Y_1 \tag{3.1a}$$

On combining equations (2.2) and (3.1) we obtain

$$a=g, \qquad e=c, \qquad f=-d, \qquad h=-b \tag{3.1b}$$

which is the necessary and sufficient condition for simple interconnection.

4. Terminations

Let us consider the doubly terminated network N with excitations E_1 and E_2 shown in Fig. 2(a). Its digital equivalent shown in Fig. 2(b) is taken in such a way that the SFD is defined to be limited between the ports 1 and 2.

Obviously

$$[X_1\ Y_1]^T=Q_1(V_1\ I_1]^T, \qquad E_1=V_1+R_sI_1 \tag{4.1a}$$

$$[X_2\ Y_2]^T=Q_2[V_2\ I_2]^T, \qquad E_2=V_2+R_LI_2 \tag{4.1b}$$

On eliminating V_1, I_1, V_2, I_2 we derive

$$X_1=(d-cR_s)^{-1}(Y_1(b-aR_s)+E_1(ad-bc)) \tag{4.2a}$$

$$X_2=(h-gR_L)^{-1}(Y_2(f-eR_L)+E_2(eh-fg)) \tag{4.2b}$$

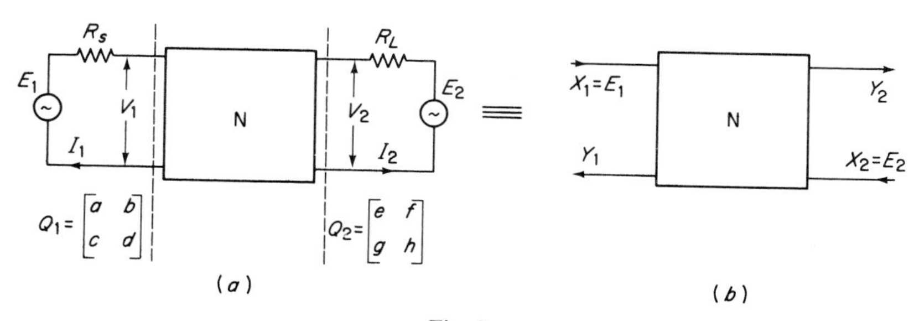

Fig. 2.

Since X_1, X_2 (Fig. 2(*b*)) are defined to be independent of Y_1, Y_2 respectively, and moreover $X_1=E_1$ and $X_2=E_2$, we have that

$$a=1, \quad b=R_2, \quad e=1, \quad f=R_L, \quad d-cR_s\neq 0, \quad h-gR_L\neq 0 \tag{4.3}$$

For the transfer function $H_d=(Y_2/E_1)_{E_2=0}$ we have

$$H_d=(h-gR_L)V_2/E_1 \tag{4.4}$$

i.e. $h-gR_L$ is a real number.

Furthermore if the reference network is reciprocal† and this property is to be maintained in the derived network, we deduce that $d-cR_s$ is also a real number, whereas for the transfer functions we have

$$(Y_2/X_1)_{X_2=0}=K(Y_1/X_2)_{X_1=0} \qquad K=(h-gR_L)/(d-cR_s) \tag{4.5}$$

5. The linearly transformed digital network as a whole.

5.1. Stability

This problem was examined in ref. 7 for wave digital filters with frequency dependent normalization port resistances. We generalize this theorem and obtain:

THEOREM 5.1. In a linearly transformed digital network the LTM of each port has elements with poles that either have strictly negative real parts or are coincident with the zeros of the voltage transfer ratio (transfer admittance) if the element is a first (second) column element of LTM respectively. The proof is similar to the one in ref. 7.

5.2. The effect of the transformation applied on the *s*-variable

The transformation (bilinear or other) applied for the transition from the analog to the digital domain imposes further constraints to the LTM elements, due to the consideration of the interconnection within the overall structure of the network.

Let us consider the network of Fig. 3, consisting of two subnetworks with modified chain matrices T_1 and T_2, simply interconnected according to equation (3.1). The choice for the LTMs of terminations is made according to equation (4.3). With reference to Fig. 3 we shall prove the following theorem.

† The network is reciprocal if and only if $(V_2/E_1)_{E_2=0}=(V_1/E_2)_{E_1=0}$.

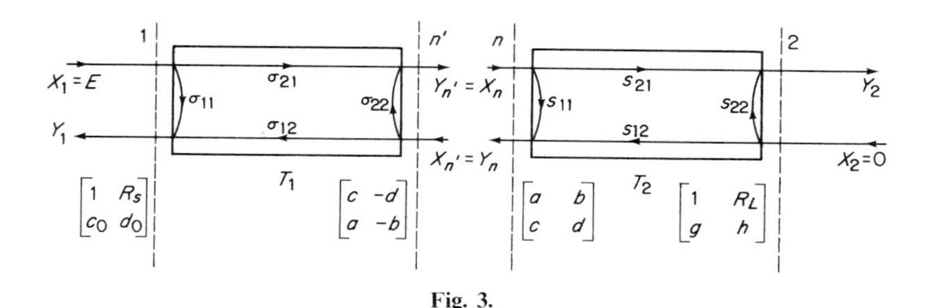

Fig. 3.

THEOREM 5.2. In every port within a SFD that devides a linearly transformed digital network in two subnetworks with modified chain matrices T_1 and T_2, at least one of the following relations holds

$$b/a \underset{s=s_0}{=} -\frac{B_1+R_sD_1}{A_1+R_sC_1} \quad (5.1a) \qquad d/c \underset{s=s_0}{=} -\frac{B_2-R_LA_2}{D_2-R_LC_2} \quad (5.1b)$$

where $s=f(z^{-1})$ is the transformation applied to the original frequency variable s, $s_0=f(0)$ and a, b, c, d are the frequency dependent elements of the LTM of the given port. R_s and R_L are the terminations of the network.

Proof: The SFD is realizable if the closed path σ_{22}, X_n, $s_{1,1}$, Y_n includes at least one delay, i.e. the product $\sigma_{22}s_{11}$ is divisible by z^{-1}. Since σ_{22} and s_{11} are rational functions z^{-1}, the function $\sigma_{22}s_{11}$ is divisible by z^{-1} if and only if it has a zero for $z^{-1}=0$. However both $\sigma_{22}(z^{-1})$ and $s_{11}(z^{-1})$ are derived from $\sigma_{22}(s)$ and $s_{11}(s)$ through the transformation $s=f(z^{-1})$, so for $z^{-1}=0$ we have $s=s_0=f(0)$. Thus the SFD is realizable if and only if the function $\sigma_{22}(s)s_{11}(s)$ has a zero for $s=s_0=f(0)$, i.e. if either $\sigma_{22}(s)=0$ or $s_{11}(s)=0$ for $s=s_0$. On recalling equation (2.2) for the two subnetworks we finally get equation (5.1).

Theorem 5.2 is of particular importance. It shows that the conditions (5.1) on the LTM elements is a consequence of the transformation $s=f(z^{-1})$ used. It also shows the free and constrained elements of LTM. Another consequence of Theorem 5.2 is that at most one of the a, b, c, d can be equal to zero, unlike other applications of linear transformations, e.g. LT active filters.[8] It can be used not only in the case of bilinear transformation but also in other transformations, e.g. digital filters imitating the behaviour of transmission line filters.[9]

Networks derived by the bilinear transformation

The case of digital filters derived from lumped reciprocal doubly resistively terminated ladder networks through the bilinear transformation $s=(1-z^{-1})/(1+z^{-1})$ should be examined in detail here. The typical SFD derived for such filters is shown in Fig. 4.

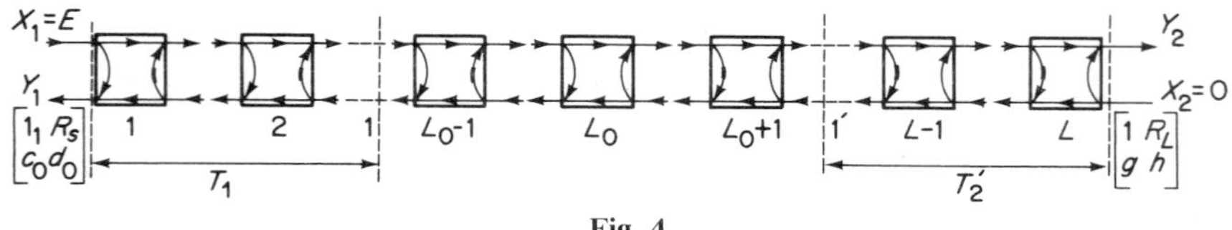

Fig. 4.

The first L_0-1 sections have σ_{22} with no delay free path whereas this holds for σ_{11} in the last $L-L_0$ sections. The L_0-section has both σ_{11} and σ_{22} delay free. The integer L_0 may vary from $L_0=1$ to $L_0=L$ (load to source and source to load procedures respectively).

It is well known that the transfer functions of passive, reciprocal, ladder networks without mutual inductances do not have zeros in the right half s-plane. On the other hand since the bilinear transformation has $s_0=f(0)=1$, the transfer function H_d of the digital filter is not divisible by z^{-1}, therefore no delay may appear in the upper path σ_{21} of the SFD. Given the reciprocity of the network this property is also found for the lower path σ_{12}.

The above shows that the described interconnection with L_0 the only parameter, is a unique way of interconnection. Indeed with reference to Fig. 4 any other interconnection requires the presence of delays in the upper or lower paths which is by hypothesis impossible.

The absence of delays in the paths σ_{21} and σ_{12} yields another property for these networks. For a given port 1 on the left of L_0, equation (5.1a) holds and moreover it is the necessary and sufficient condition for the presence of delays in the σ_{22} paths of not only the overall network on the left of 1, but also of each constituent network block lying on the left of 1. If the port 1 lies on the right of L_0 the corresponding necessary and sufficient condition is given by equation (5.1(b). It effects the presence of delays in the σ_{11} paths of all the network blocks lying on the right of port 1.

From the above it is evident that for the ports on the left of L_0, d and c are free parameters and only the ratio b/a is constrained for $s=1$. On the contrary for ports on the right of L_0, a and b are free parameters and only the ratio d/c is constrained for $s=1$.

6. Conclusions

The properties of linear transformations were examined. They describe explicitly the behaviour of LTs in digital filter design. It was found that the bilinear transformation used for the transition from the analog to digital domain imposes a constraint that limits our degree of freedom in the design. This constraint effects on the one hand the values of the multipliers within the filter and on the other the simplicity of the derived structures. However, other criteria for simple structure realization remain to be investigated. It is evident that the results derived include all the subcases of linear transformations described in the literature e.g. wave digital filters, I.V.R. and M.T.A. transformations, etc.

References

1. A. Fettweis, "Digital Filter Structures Related to Classical Filter Networks", *AEU*, **25**, pp. 78–89 (1971).
2. A. Fettweis, "Pseudopassivity, Sensitivity, and Stability of Wave Digital Filters", *IEEE Trans. on Circuit Theory*, **CT-19**, pp. 668–673 (1972).
3. A. Sedlmeyer and A. Fettweis, "Digital Filters with True Ladder Configuration", *Int. Jour. on Circuit Theory and Applic.*, **1**, pp. 5–10 (1973).
4. A. G. Constantinides, "Alternative Approach to Design of Wave Digital Filters", Electronics Letters, **10**, pp. 59–60 (1974).
5. S. S. Lawson and A. G. Constantinides, "A Method for deriving Digital Filter Structures from Classical Filter Networks", *Proc.* 75 *ISCAS*.
6. A. G. Constantinides, "Design of digital filters from LC ladder networks", *Proc. IEEE* **123**, No 12 (Dec. 1976).
7. E. Tzanettis and A. G. Constantinides, "Wave Digital Filters with Frequency Dependent Normalization Port Resistances", 1978 European Conf. on Circuit Circuit Theory and Design.
8. H. Dimopoulos, "Design of Electronic Active Filters from Ladder Networks using Linear Transformations", Ph.D. Thesis, University of London, 1978.
9. S. S. Lawson, "Digital Filter Structures from Classical Analogue Networks", Ph.D. Thesis, University of London, 1975.

Generalized Orthogonal Filters for Stochastic Prediction and Modelling

E. DEPRETTERE and P. DEWILDE

Network Theory Section, Department of Electrical Engineering, Delft University of Technology, Delft, The Netherlands

1. Introduction

From stochastic prediction and modelling theory, it is known[1] that for a stationary gaussian process $\{y_k\}$ characterized by its "positive" covariance $\{r_k\}$, a least-squares optimal linear predictor can be recursively build up from the covariance data by the Levinson algorithm.[2] The successive predictors of increasing order are orthogonal polynomials.[3,4] The inverse nth order predictor is an autoregressive (AR) or all-pole model, it produces from white noise a process $(\hat{y}_k\}$ whose covariance $\{\hat{r}_k\}_n$ is an nth order approximation of $\{r_k\}$. The (eventually estimated) covariance sequence $\{r_k\}$ with $r_{-k}=\tilde{r}_k$($\sim$is the hermitian conjugate) has the property that the one-side z-transform

$$Z_0 \triangleq r_0+2r_1z^{-1}+2r_2z^{-2}+\dots \tag{1}$$

is analytic outside the unit disc with positive real part there. Let

$$\tfrac{1}{2}[\tilde{Z}_0(e^{j\theta})+Z_0(e^{j\theta})]=T_A(e^{j\theta})\tilde{T}_A(e^{j\theta})=\tilde{T}_B(e^{j\theta})T_B(e^{j\theta}) \tag{2}$$

be a spectral factorization of the density function

$$R(\theta)=\tfrac{1}{2}[\tilde{Z}_0(e^{j\theta})+Z_0(e^{j\theta})] \tag{3}$$

then any filter with transfer function T_A is a "modelling" filter which, when inputed with white noise will produce the covariance Z_0, while any filter with transfer function T_A^{-1} will be a "prediction" filter which, when inputed with a signal with the density (3) will output white noise. One nice way to realize such filters is by means of an orthogonal filter structure†.[5]

---

† By an orthogonal filter structure we mean a compatible cascade structure of elementary sections each of which consists of a minimal lossless dynamic portion (e.g. a single delay) and a unitary connecting part.

By a compatible structure we mean a deadlock-free structure, which in the underlying case implies flow-diagrams without delay-free loops.

Orthogonal filters are generalized wave digital filters[6,14] and may be described either by means of a scattering matrix $\Sigma(z)$:

$$\begin{bmatrix} B_1 \\ A_2 \end{bmatrix} = \begin{bmatrix} \Sigma_{11} & \Sigma_{12} \\ \Sigma_{21} & \Sigma_{22} \end{bmatrix} \begin{bmatrix} A_1 \\ B_2 \end{bmatrix} \tag{4}$$

or by means of a chain scattering matrix (CSM) $\Theta(z)$†:

$$\begin{bmatrix} A_1 \\ B_1 \end{bmatrix} = \begin{bmatrix} \Theta_{11} & \Theta_{12} \\ \Theta_{21} & \Theta_{22} \end{bmatrix} \begin{bmatrix} A_2 \\ B_2 \end{bmatrix}, \quad \Theta = \begin{bmatrix} \Sigma_{21}^{-1} & 0 \\ 0 & \Sigma_{12*}^{-1} \end{bmatrix} \begin{bmatrix} I & -\Sigma_{22} \\ -\Sigma_{22*} & I \end{bmatrix} \tag{5}$$

which satisfy contractivity (passivity) and unitary (losslessness) conditions as discussed in the literature:[5,6]

(1) $\tilde{\Sigma}\Sigma \leqq I \qquad (\tilde{\Theta} J \Theta \geqq J) \quad \text{in } |z| \geqq 1$
(2) $\Sigma_* \Sigma = I \qquad (\Theta_* J \Theta = J)$

If the filterports are connected to sources E_i as shown in Fig. 1, then one may define a forward transfer gain T_A:

$$A_2(z)|_{e_2=0} = T_A(z) E_1(z)$$

and a backward transfer gain T_B:

$$B_1(z)|_{e_1=0} = T_B(z) E_2(z)$$

With these transfer gains, equation (5) can be written as:

$$\Theta(z) = \begin{bmatrix} T_A^{-1}(Z_0+1)/2 & -T_A^{-1}(Z_0-1)/2 \\ -T_{B*}^{-1}(Z_{0*}-1)/2 & T_{B*}^{-1}(Z_{0*}+1)/2 \end{bmatrix} \tag{6}$$

where $Z_0(z) \triangleq (1+\Sigma_{22})(1-\Sigma_{22})^{-1}$ is positive. The losslessness of the filter ensures that relation (2) is satisfied. Hence it follows that a lossless digital filter will produce a modeling and a prediction filter if it is so designed that:

(1) $[1+\Sigma_{22}][1-\Sigma_{22}]^{-1}$ is Z_0 or at least approximately Z_0,
(2) it is either minimal phase or maximal phase.

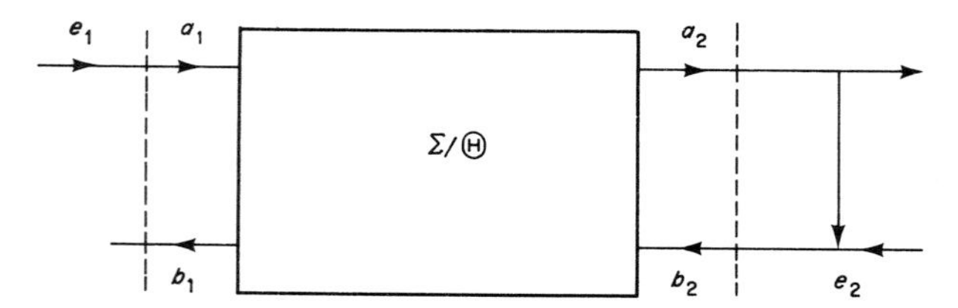

Fig. 1. A Σ/Θ digital filter.

† $\Sigma_*(z) \triangleq \tilde{\Sigma}(\bar{z}^{-1})$; $\quad J \triangleq \begin{bmatrix} 1 & 0 \\ 0 & -1 \end{bmatrix}$.

In the next paragraph we shall summarize a classical algorithm which produces an orthogonal structure to approximate a desired covariance sequence Z_0.

2. The Levinson algorithm in the wave digital context

We give a quick review of the Levinson theory[2,3,7] and connect it with our own needs. Starting from a covariance sequence $\{r_k\}$, the Levinson algorithm recursively computes polynomials $\phi_n(z)$, $\psi_n(z)$ which are orthogonal on the unit circle by means of the formulas:

$$\begin{bmatrix} \phi_n(z) \\ \phi_n^*(z) \end{bmatrix} = \frac{1}{\sqrt{1-|k_n|^2}} \begin{bmatrix} z & \varepsilon k_n \\ \varepsilon k_n z & 1 \end{bmatrix} \begin{bmatrix} \phi_{n-1}(z) \\ \phi_{n-1}^*(z) \end{bmatrix} \tag{7}$$

where

(i) $\phi_n = \begin{cases} \phi_n \leftarrow \varepsilon = -1 \\ \psi_n \leftarrow \varepsilon = +1 \end{cases}$

(ii) $\phi_n^*(z) \triangleq z^n \bar{\phi}_n(\bar{z}^{-1})$

(iii) k_n is called the nth reflection coefficient and can be computed from $\{r_k\}$. These coefficients have the property that $|k_n| < 1$, all n,

(iv) the polynomials ϕ_n are orthogonal with respect to the density function (3).

The matrix

$$\theta_\eta(z) = \frac{1}{\sqrt{1-|k_\eta|^2}} \begin{bmatrix} z & -k_\eta \\ -k_\eta z & 1 \end{bmatrix} \tag{8}$$

defines a chain scattering matrix factor which is lossless in our former definition. The two recursions (7) then produce a global chain scattering matrix:

$$\Theta_n(z) = \theta_n(z) \ldots \theta_\eta(z) \ldots \theta_1(z) = \frac{1}{2} \begin{bmatrix} \phi_n + \psi_n & \phi_n - \psi_n \\ \phi_n^* - \psi_n^* & \phi_n^* - \psi_n^* \end{bmatrix} \tag{9}$$

satisfying the same passivity properties. It is known[3,8] that if n is large enough, then:

(i) $\psi_n/\phi_n = Z_{0n} \approx Z_0$ (10a)

(ii) $1/\phi_n = T_{An} \approx T_A$ (10b)

so that the CSM (9) defines an autoregressive (AR) approximation[7] to the desired covariance Z_0.

The cascade (9) can be used as a prediction filter with transfer function

$$T_A^{-1} = \phi_n/z^n$$

by realizing the unnormalized product (Fig. 2*a*) defined by:

$$\begin{bmatrix} \phi_n/z^n \\ \phi_n^*/z^n \end{bmatrix} = \begin{bmatrix} 1 & -k_n z^{-1} \\ -k_n & z^{-1} \end{bmatrix} \cdots \begin{bmatrix} 1 & -k_\eta z^{-1} \\ -k_\eta & z^{-1} \end{bmatrix} \cdots \begin{bmatrix} 1 & -k_1 z^{-1} \\ -k_1 & z^{-1} \end{bmatrix} \begin{bmatrix} 1 \\ 1 \end{bmatrix} \tag{11}$$

The corresponding modelling filter is obtained by realizing the Σ-filter represented by equation (9). Indeed, comparing equations (9) and (10*a*, *b*) with equation (6) it follows that the filter of Fig. 1 with $e_2 = 0$ has a transfer function $T_A = 1/\phi_n$. Using the defining relations (4) and (5), the structure of the modelling filter is immediately obtained by a graphical trick on Fig. 2(*a*). It is depicted in Fig. 2(*b*)†. Its transfer function is $T_A = z^n/\phi_n$ and its output covariance (when inputed with gaussian white noise) is Z_{0n}. The equal structures (undirected graphs) of the filters in Fig. 2 and the computability[5] of the modelling flow diagram (directed graph) result from the direct path between input and output and from the delays between the lattice filter sections respectively.

The realization of $T_A = z^n/\phi_n$ by means of an orthogonal structure[5] (wave digital filter) is attractive since the coefficients k_η have better numerical properties than the parameters of ϕ_n. Moreover, the k_η's may have physical meaning, see e.g. ref. 2, as reflection coefficients in a segmented acoustic tube model of the human vocal tract. Thus, instead of using the Levinson recursion producing ϕ_n, a direct recursive construction of the cascade (11), producing k_η, may be preferable. This can be achieved‡ by a Darlington recursion[10,11] as follows:

$$\frac{1}{z^n}\begin{bmatrix} -\Delta_{2*}^{(0)} \\ \Delta_{1*}^{(0)} \end{bmatrix}^T = \begin{bmatrix} -\Delta_{2*}^{(n)} \\ \Delta_{1*}^{(n)} \end{bmatrix}^T \overset{1}{\underset{\mu=n}{\overset{\curvearrowleft}{\prod}}} \begin{bmatrix} 1 & -k_\mu \\ -k_\mu & 1 \end{bmatrix} \begin{bmatrix} I_2 + \frac{z-1}{z} x_\mu \tilde{x}_\mu J \end{bmatrix} \tag{12}$$

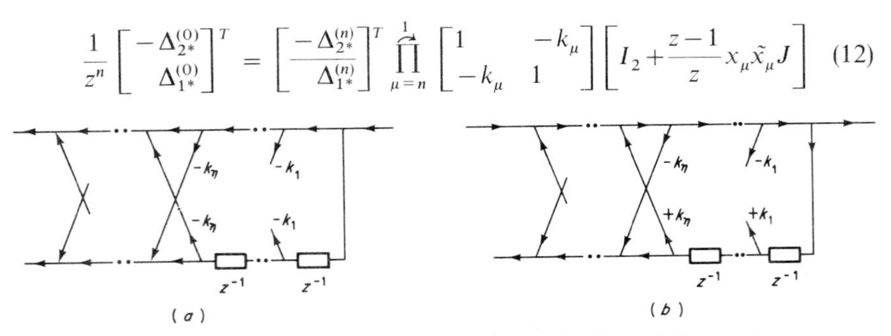

Fig. 2. Levinson–lattice filter: (*a*) prediction mode; (*b*) modelling mode.

---

† Variants to this structure can be found in ref. 2.

‡ A very efficient alternative in which both the covariance and the reflection coefficients are recursively estimated from time domain data is known as Burg's algorithm.[9]

where

(1) $$\Delta_{.*}^{(m)}(z) \triangleq u(z)d^{(m)}_{.} = [1\ z\ z^2\ ..] \begin{bmatrix} d.(0) \\ d.(1) \\ . \\ . \end{bmatrix}^{(m)} \qquad (m=1,..,n)$$

(2) $$\Delta_{2*}^{(0)} = \frac{1}{2}(Z_{0*}-1), \qquad \Delta_{1*}^{(0)} = \frac{1}{2}(Z_{0*}+1)$$

(3) $$\frac{1}{z}\left[-\Delta_{2*}^{(\mu-1)}\ \Delta_{1*}^{(\mu-1)}\right] = \left[-\Delta_{2*}^{(\mu)}\ \Delta_{1*}^{(\mu)}\right]\begin{bmatrix} 1 & -k_\mu \\ -k_\mu & 1 \end{bmatrix}\left[I_2 + \frac{z-1}{z}x_\mu \tilde{x}_\mu J\right]$$

(4) $$|\Delta_{2*}^{(m)}[\Delta_{1*}^{(m)}]^{-1}| < 1, \quad \text{all m},$$

(5) $$x_\mu \sim [\Delta_{2*}^{(\mu-1)}(0)\ \Delta_{1*}^{(\mu-1)}(0)]^T, \quad \tilde{x}_\mu J x_\mu = -1.$$

If we further require the resulting cascade to be computationally compatible† [5], then $d_{20}^{(m)}(m=1,\ldots,n)$ must be zero whence $x_{\mu 1}=0$, all μ. We then have the following.

Algorithm. Reflection coefficients k_μ of the L^2- filter.

$d_1^{(.)}(0)=1;$
$k_1 := d_2^{(0)}(1);$

for $1 \leftarrow \mu$ upto n begin
for $\mu \leftarrow \gamma$ upto 1 begin

$$\begin{bmatrix} d_1(\gamma) \\ d_2(\gamma) \end{bmatrix} := \left\{\frac{1}{1-k^2}\begin{bmatrix} 1 & -k \\ -k & 1 \end{bmatrix}\right\}_{\mu-\gamma+1}\begin{bmatrix} d_1(\gamma) \\ d_2(\gamma+1) \end{bmatrix}; \tag{13}$$

end;

$k_{\mu+1} = -d_2(1)$; end; end; □

The major drawback of this classical polynomial predictor method is that it succeeds only in producing autoregressive (all-pole) transfer functions. The generalization of the method to autoregressive-moving-average (ARMA) or pole-zero filters is discussed in the next paragraph.

3. Generalization of the orthogonal approximate predictive recursion to rational predictors and ARMA models

The recursive extraction of factors

$$\begin{bmatrix} 1 & -kz^{-1} \\ -k & z^{-1} \end{bmatrix} \tag{14}$$

† Recall that compatibility implies noniterative computations. As shown in ref. 5, this condition requires that $\Delta_2^{(.)}(0)$ must be zero.

from

$$\frac{1}{z^n}\left[-\frac{1}{2}(Z_{0*}-1)\ \frac{1}{2}(Z_{0*}+1)\right]$$

in a factorization (12) produces a structure with transmission zeros only at $z=0$, hence an AR-model. However, if other transmission zeros are desired, an efficient model will be of the ARMA-type and the same method[10] can be used since it is possible to extract factors at other points inside the unit circle, now from

$$\frac{1}{z^r}\left(\Pi\frac{1}{z-a_\mu}\right)\left[-\frac{1}{2}(Z_{0*}-1)\ \frac{1}{2}(Z_{0*}+1)\right] \tag{15}$$

The resulting cascade structure will now contain r factors with transmission zero at zero and $n-r$ factors with transmission zero at a_μ, $\mu=1, \ldots, (n-r)$, $|a_\mu|<1$, a typical one being (for real a):

$$\begin{bmatrix} 1 & -k\dfrac{1-az}{z-a} \\ -k & \dfrac{1-az}{z-a} \end{bmatrix} \tag{16}$$

Obviously, for $a=0$, expression (16) reduces to expression (14). Denoting $\xi_a=(z-a)/(1-az)$ it may be observed that the generalized chain scattering factor is similar to the Levinson factor (14), so that at first glance one may be tempted to conclude that the generalized filters are likewise represented by the flow graphs of Fig. 2, with the understanding that the interleaving shifts are replaced by the nonuniform delay operators $\xi_{a\mu}$ of which a flow diagram is depicted in Fig. 3.

Unfortunately, this particular structure of the ARMA-modelling filter is not computationally compatible in general, which is due to the instantaneous mutual dependence of forward and backward travelling signals. However, we have shown elsewhere[5] that this computability problem which normally occurs in Σ/Θ-filters is solvable in general. The compatible factors can be extracted directly from expression (15) by means of a Darlington

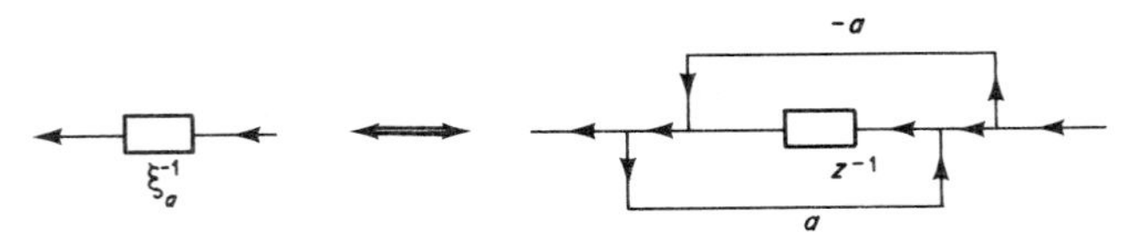

Fig. 3. Generalized nonuniform delay element ξ_a^{-1}.

recursion which generalizes (12), as follows:

$$\frac{1}{z^r}\left(\Pi\frac{1-a_\eta z}{z-a_\eta}\right)\begin{bmatrix}-\Delta_{2*}^{(0)}\\ \Delta_{1*}^{(0)}\end{bmatrix}^T=\begin{bmatrix}-\Delta_{2*}^{(n)}\\ \Delta_{1*}^{(n)}\end{bmatrix}^T\overset{1}{\curvearrowleft}\prod_{\mu=n} K_\mu\begin{bmatrix}1 & k'_\mu\\ k'_\mu & 1\end{bmatrix}\times\left[I_2+\frac{z-1}{(1-a_\mu)(z-a_\mu)}x_\mu\tilde{x}_\mu J\right] \tag{17}$$

where

(1) $\Delta_{2*}^{(m)}(0)=0,\quad \Delta_{1*}^{(m)}(0)=1\quad (m=1,\ldots,n)$

(2) $$\frac{1-a_\mu z}{z-a_\mu}[-\Delta_{2*}^{(\mu-1)}\quad \Delta_{1*}^{(\mu-1)}]=[-\Delta_{2*}^{(\mu)}\quad \Delta_{1*}^{(\mu)}]K_\mu\begin{bmatrix}1 & k'_\mu\\ k'_\mu & 1\end{bmatrix}\left[I_2+\frac{z-1}{(1-a_\mu)(z-a_\mu)}x_\mu\tilde{x}_\mu J\right]$$

(3) $|\Delta_{2*}^{(m)}[\Delta_{1*}^{(m)}]^{-1}<1, |z|<1$, all m

(4) $x_\mu=\left(\frac{1-a_\mu^2}{1-|k|_\mu^2}\right)^{k/2}[k_\mu-1]^T,\ -k_\mu=\Delta_{2*}^{(\mu-1)}(a_\mu)/\Delta_{1*}^{(\mu-1)}(a_\mu)$

As in the polynomial case, K_μ and k'_μ are determinated by the requirement that

$$[-\Delta_2^{(\mu)}(0)\quad \Delta_1^{(\mu)}(0)]\equiv[0\quad 1]\qquad (\text{all }\mu)$$

by which computability is ensured. It is worthwhile to remark that it has been shown[8] for stochastic modelling filters that keeping the cascade deadlock-free ensures the optimality in the least-squares sense as well as the convergence of the recursive model. The latter then generalizes the known convergence property (equation (10*b*)) of polynomial models. Note that r out of the n factors have $a=0$ and $k=0$. They can be extracted by means of the algorithm given in the preceeding paragraph which will appear to be the limiting case of a similar non-zero-pole extraction algorithm.

In case $a\neq 0$, it can be shown[5] that

$$\theta(z;\ x,\ a)=K\begin{bmatrix}1 & k'\\ k' & 1\end{bmatrix}\left[I_2+\frac{z-1}{(1-a)(z-a)}\ x\ \tilde{x}\ J\right]$$

is equivalent to

$$\frac{J}{1+|y_1|^2}\left\{\begin{bmatrix}1 & 0\\ 0 & a\end{bmatrix}+\frac{y\tilde{y}J}{z-a}\begin{bmatrix}z & 0\\ 0 & 1\end{bmatrix}\right\} \tag{18}$$

and also to

$$\frac{1}{1-|k|^2}\begin{bmatrix}1 & -(k/a)\xi_a^{-1}\\ -(k/a) & \xi_a^{-1}\end{bmatrix}\begin{bmatrix}1 & -k\\ -k & 1\end{bmatrix} \tag{19}$$

where

(1) $$y = \frac{1}{(1-|x_2|^2)^{1/2}} \begin{bmatrix} 1 & 0 \\ 0 & a \end{bmatrix} x = \left(\frac{1-a^2}{1-k^2/a^2} \right)^{1/2} [k/a - 1]^T$$

(2) $|a| > |k|$ by Schwarz's Lemma[12]

(3) $$\xi_a^{-1} = \frac{z^{-1} - a}{1 - a\, z^{-1}} \tag{20}$$

For the extraction of a single factor $\theta(z;k,a)$ we then have the following:

Algorithm. Extraction of a transmission zero at a.

$d_2(0) = 0$; $d_1(0) = 1$;

$k = -\Delta_{2*}(a)/\Delta_{1*}(a)$;

for $1 \leftarrow l$ upto L begin

$$\begin{bmatrix} d_1(l) \\ d_2(l) \end{bmatrix} := \frac{1}{1-k^2/a^2} \begin{bmatrix} 1 & k/a \\ k/a & 1 \end{bmatrix} \begin{bmatrix} d_1(l) + kd_2(l) \\ \delta_2(l) + k\delta_1(l) \end{bmatrix}; \tag{21}$$

end; end;†
where $u(z)\delta = B^{-1}\{(B(u(z)d) - u(a)d)\xi_a^{-1}\}$, B being the bilinear argument transform (20) mapping the unit disc onto itself, fixing the point 1. □

The flow graphs corresponding to the elementary rational prediction and ARMA modelling factor (18) are depicted in Fig. 4. Both diagrams have again identical undirected graphs. The modelling section will not cause deadlock situations when inserted in a cascade. Moreover, it not only realizes a stable invertible ARMA-type transfer function (a minimal phase

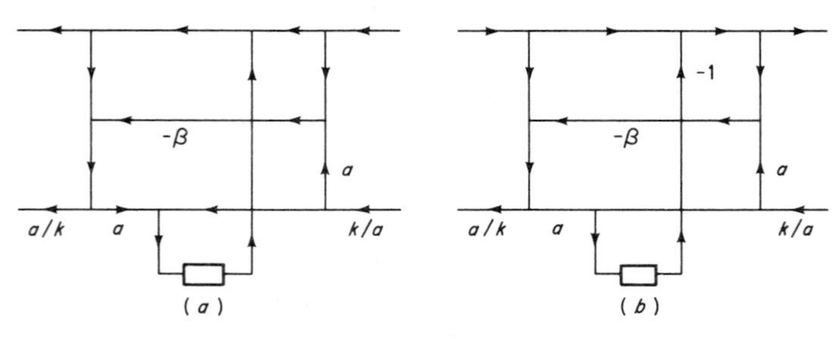

Fig. 4. Factor with nonzero pole: (*a*) predictor graph; (b) model graph: $|a| < 1$, $|\frac{k}{a}| < 1$, $\beta = (1 - k/a^2)(1 - k^2)^{-1}$.

† If $a = 0$, then $k = 0$, $k/a \Longrightarrow -k$ and $\delta_2(l) = d_2(l+1)$ so that expression (21) reduces to (13).

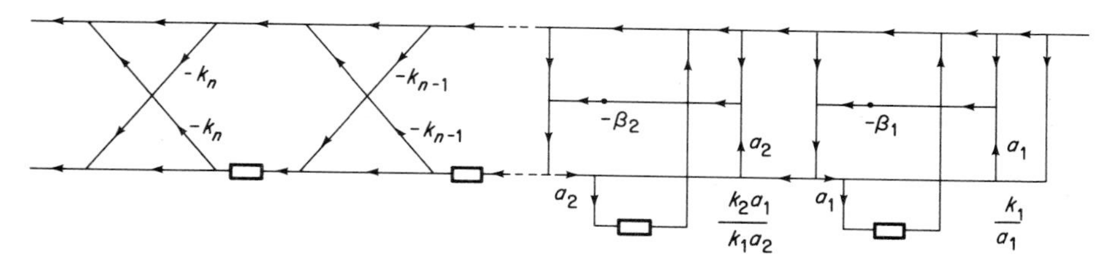

Fig. 5. General rational predictor cascade.

function) it also provides for a stable implementation due to the orthogonality of the computations.[5]

It is interesting to remark that a general prediction filter as represented in Fig. 5 is equivalent to that of Fig. 6, by applying the following identities: (1) Using (19):

$$\frac{1}{1-k^2}\begin{bmatrix}1 & -(k/a)\xi_a^{-1}\\ (k/a) & \xi_a^{-1}\end{bmatrix}\begin{bmatrix}1 & -k\\ -k & 1\end{bmatrix}\begin{bmatrix}1\\ 1\end{bmatrix}=\begin{bmatrix}1 & -\rho z^{-1}\\ -\rho & z^{-1}\end{bmatrix}\begin{bmatrix}1/(1-az^{-1})\\ 1/(1-az^{-1})\end{bmatrix}$$

$$\rho=\frac{a+(k/a)}{1+a(k/a)}, \qquad |\rho|<1 \tag{22}$$

(2) Using (19) and (14):

$$\frac{1}{1-k^2}\begin{bmatrix}1 & -(k/a)\xi_a^{-1}\\ -(k/a) & \xi_a^{-1}\end{bmatrix}\begin{bmatrix}1 & -k\\ -k & 1\end{bmatrix}\begin{bmatrix}1 & -lz^{-1}\\ -l & z^{-1}\end{bmatrix}=\begin{bmatrix}1 & -\boldsymbol{l}z^{-1}\\ -\boldsymbol{l} & z^{-1}\end{bmatrix}$$

$$\times\frac{1}{1-\boldsymbol{k}^2}\begin{bmatrix}1 & -(k/a)\xi^{-1}\\ -(\boldsymbol{k}/a) & \xi_a^{-1}\end{bmatrix}\begin{bmatrix}1 & -\boldsymbol{k}\\ -\boldsymbol{k} & 1\end{bmatrix} \tag{23}$$

where

$$\boldsymbol{l}=\frac{(k/a)(1-a^2)-al(1-(k/a)^2)}{1-a^2(k/a)^2}$$

$$l=\frac{(\boldsymbol{k}/a)(1-a^2)-a\boldsymbol{l}\,(1-(\boldsymbol{k}/a)^2)}{1-a^2(\boldsymbol{k}/a)^2}$$

$$(\boldsymbol{k}/a)=\frac{l+a(k/a)}{1+al(k/a)}, \qquad (k/a)=\frac{\boldsymbol{l}+a(\boldsymbol{k}/a)}{1+a\boldsymbol{l}(\boldsymbol{k}/a)}$$

$$\frac{1-(\boldsymbol{k}/a)^2a^2}{1+(\boldsymbol{k}/a)a\boldsymbol{l}}=\frac{1-(k/a)^2a^2}{1+(k/a)al}$$

$$\frac{1-k^2}{1-\boldsymbol{k}^2}=\left[\frac{(1-(k/a)^2)(1-k^2)(1-l^2)}{(1-(\boldsymbol{k}/a)^2)(1-\boldsymbol{k}^2)(1-\boldsymbol{l}^2)}\right]^{1/2}$$

The equivalent structure of Fig. 6 is not surprising since it consists of a

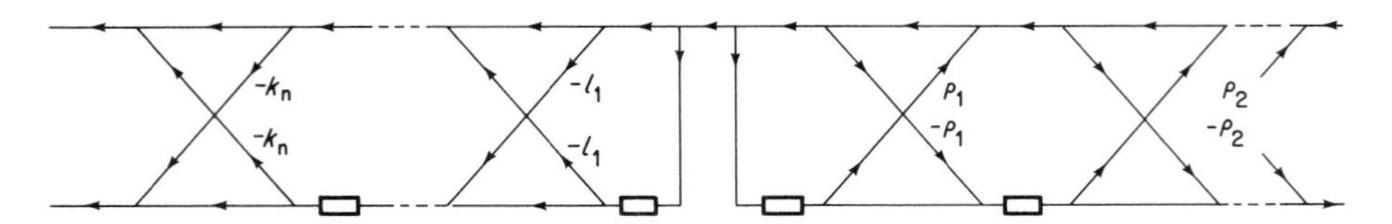

Fig. 6. Prediction filter equivalent to that of Fig. 5 showing the autoregressive Σ-prefilter.

cascade of an all-pole (AR) and an all-zero (MA) part corresponding to the denominator and the numerator respectively of the inverse ARMA transfer function. The autoregressive right-hand portion can be viewed as a prefilter which converts the ARMA-function to an AR-function thereby relating the original problem to the specialized Levinson-type problem.

4. Conclusion

We have shown that the known lattice filter realization of the stable invertible polynomial predictors produced by the classical Levinson algorithm is a "backward" used unit element type wave digital filter which realizes the inverse polynomial. This result has been used to derive a recursive approximate predictive synthesis procedure. The filter construction method has then been generalized to produce ARMA-modelling filters as well. These generalized prediction and modelling filters are still characterized by a common filter structure which has an orthogonal and deadlock-free implementation. The construction of generalized orthogonal cascade structures for stochastic filtering has its counterpart in interpolation theory, where the solution of the Nevanlinna-Pick problem[13] leads to triangularization and decomposition of the so-called Pick matrix.

The transformation by which a Pick matrix can be converted to a Toeplitz matrix[13] can be interpreted in terms of the equivalence we mentioned in our last paragraph between the minimal orthogonal ARMA-filter on the one hand and the cascade of a wave digital AR-filter and a reverse wave digital MA-filter on the other.

References

1. J. Makhoul, "Linear Prediction: A tutorial review", *Proc. IEEE*, **62**, (1975).
2. J. Markel and A. Gray Jr., "Linear Prediction of Speech". Berlin: Springer-Verlag, 1976.
3. L. Ya. Geronymus, "Orthogonal Polynomials", Transl. Consult. Bur. New-York, 1961.
4. Ph. Delsarte, Y. Genin and Y. Kamp, "Orthogonal Polynomial Matrices on the Unit Circle", *MBLE Report R* 349 (1977).
5. E. Deprettere, and P. Dewilde, "Orthogonal Cascade Realization of Real Multiport Digital Filters" *Int. J. Circ. Theor. Appl.* (1980) to appear.

6. R. Newcomb, "Linear Multiport Synthesis". New-York: McGraw-Hill, 1966.
7. Th. Kailath, "A View of Three Decades of Linear Filtering Theory", *IEEE Trans. on Inf. Theory*, **IF-20** (1976).
8. P. Dewilde, "On the Convergence of the Generalized Szegö-Levinson Least-Square Error Algorithm", Int. Symp. on Information Theory, Grignano, Italy, 1979.
9. J. P. Burg, "Maximum entropy spectral analysis", 37th Annual Meeting Society Explor. Geophys., Oklahoma City, OK, 1967.
10. P. Dewilde, A. Vieira and Th. Kailath, "On the generalized Szegö-Levinson Realization Algorithm for Optimal Linear Predictors based on a Network Synthesis Approach", *IEEE Trans. on Circ. and Systems*, **CAS-25**(1978).
11. E. Deprettere "A Numerical Calculus for Constructing Approximating Covariance Prediction/Modeling Filters in the Time Domain", ECCTD'78 Conference, Lausanne, 1978.
12. W. Rudid, "Real and Complex Analysis". New York: McGraw-Hill, 1974.
13. Ph. Delsarte, Y. Genin and Y. Kamp, "The Nevanlinna–Pick problem for matrix valued functions", *MBLE Report R* 366 (1978).
14. A. Ali, M., A. G. Constantinides, "Design of Low Sensitivity and Complexity Digital Filter Structures", ECCTD'78 Conference, Lausanne, 1978.

Two-dimensional Phase Filtering

G. GARIBOTTO

Centro Studi e Laboratori Telecomunicazioni S.p.A, Torino, Italy

Introduction

The advantages of using two-dimensional recursive filters instead of finite-impulse-response operators are mainly due to a saving in computing time and memory requirements, and to an increased design flexibility related to their infinite impulse response behaviour. On the other hand, due to the difficulty in designing a stable two-dimensional recursive filter, an implementation which accomplishes that potential performance must be guaranteed. Among the various designing techniques which have been proposed in the last few years two different approaches can be distinguished according to the chosen structure of the system.

In the first case[1] a cascade form of causal filters is accepted, where the output is evaluated in real time, with minimum delay due to the order of the difference equation and to the group delay of the overall system. The memory requirement is given by the state matrices of the filters connected in cascade. The impulse responses of such filters can be defined either in a quadrant of the space domain or in the half-plane,[2,3] which allows better performance in magnitude and phase response behaviour. These filters are mainly designed as a product of one-dimensional systems,[4] or by using nonlinear techniques[1] or with linearized procedures.[5]

According to a different strategy, a noncausal structure is accepted, by using a cascade of filters with different directions of recursion.[4] Such filters are usually implemented by rotating through multiples of $\beta = 90°$ the output matrix of each section in cascade; this requires a large storage of data and an increased delay, in addition to the typical constraints of a two-dimensional recursive filter. As a consequence of an IIR implementation with respect to an FIR is no longer advantageous.

In this paper the first approach is considered, in which case the phase response is particularly important, being responsible for the impulse response behaviour of the system.

In section 1 the mathematical model of the filter is developed and some useful design techniques are described, according to amplitude as well as phase frequency specifications. Some typical applications of two-dimensional phase filtering are then discussed, starting with an all-pass phase equalizer design.

Another example is devoted to an application in the field of seismic signal processing where a phase filter is required to model the wave propagation through the earth's layers. Finally minimum-phase filters are used in the field of image processing.

1. Mathematical model

In the following a two-dimensional recursive filter is represented by a general transfer function:

$$H(z_1, z_2)=\prod_{r=1}^{R} H_r(z_1, z_2)\prod_{s=1}^{S} H_{E_s}(z_1, z_2) \tag{1}$$

where $H_r(z_1, z_2)$ is a ratio of polynomials determining the amplitude response of the digital filter:

$$H_r(z_1, z_2)=\frac{P_r(z_1, z_2)}{Q_r(z_1, z_2)}=\frac{\sum_{m=0}^{N}\sum_{n=0}^{N} p_r(m, n)z_1^m z_2^n}{\sum_{m=0}^{N}\sum_{n=0}^{N} q_r(m, n)z_1^m z_2^n} \tag{2}$$

$H_{E_s}(z_1, z_2)$ represents an all-pass phase filter with a transfer function:

$$H_{E_s}(z_1, z_2)=\frac{z_1^N z_2^N D_s(z_1^{-1}, z_2^{-1})}{D_s(z_1, z_2)}=\frac{\sum_{m=0}^{N}\sum_{n=0}^{N} d_s(N-m, N-n)z_1^m z_2^n}{\sum_{m=0}^{N}\sum_{n=0}^{N} d_s(m, n)z_1^m z_2^n} \tag{3}$$

and a phase response

$$<H_{E_s}(\mathrm{e}^{-j\omega_1}, \mathrm{e}^{-j\omega_2})=-N\omega_1-N\omega_2-2<D_s(\mathrm{e}^{-j\omega_1}, \mathrm{e}^{-j\omega_2}) \tag{4}$$

The filter structure (1) must be consistent with the particular problem in question and may contain a cascade of amplitude filters (2), phase filters (3) or both. Moreover, according to the chosen design procedure the filter structure (1) can be considered as a whole or each term in cascade can be designed in an iterative manner.

2. Filter design

Due to the great importance of phase in two-dimensional signal processing, filter design techniques must be developed which use phase as well as magnitude frequency specifications.

2.1. Complex frequency specifications

In a first approach the coefficients of the filter can be determined with a minimization of an appropriate functional depending on the deviation from a prescribed complex frequency response. This can be done by using either nonlinear techniques[1,6] or an iterative linear method[5] where the nonlinear terms, in the error function to be minimized, are previously evaluated, starting from a given initial condition, so that a linear problem is solved at each iteration. This method is briefly described in the following.

Let a complex frequency function

$$F(\omega_1, \omega_2) = |F(\omega_1, \omega_2)| e^{j\phi(\omega_1, \omega_2)}$$

be given. It can be separated in a real and imaginary part and evaluated at fixed frequency pairs $(\omega_{1_k}, \omega_{2_k})$ which, in a shorter notation, results in

$$F_k = R_k + jI_k = |F_k| \cos \phi_k + j|F_k| \sin \phi_k \tag{5}$$

this function has to be approximated in a certain frequency interval by a cascade of two-dimensional recursive filters like (2), whose coefficients must be optimized, each section at a time. The frequency response of such a filter, evaluated at the fixed pairs $(\omega_{1_k}, \omega_{2_k})$ can be written in alternative ways:

$$H_k = P_k / Q_k = |H_k| e^{j\theta_k} = X_k + jY_k \tag{6}$$

The error function to be minimized is:

$$\varepsilon = \sum_{k=1}^{K} |\varepsilon_k|^2 = \sum_{k=1}^{K} |F_k - \frac{P_k}{Q_k}|^2 = \sum_{k=1}^{K} \{(R_k - X_k)^2 + (I_k - Y_k)^2\} \tag{7}$$

Due to the rational function of the filter, this is a typical nonlinear problem which can be converted into a linear form by using a weighted error function in an iterative way,[7] so that at the step i:

$$E_1 = \sum_{k=1}^{K} |\varepsilon_k Q_{k,i} / Q_{k,i-1}|^2 = \sum_{k=1}^{K} |(F_k Q_{k,i} - P_{k,i}) / Q_{k,i-1}|^2 \tag{8}$$

where $Q_{k,i-1}$ represents the value at frequencies $(\omega_{1_k}, \omega_{2_k})$ of the denominator, whose coefficients have been determined at the previous step.

The necessary condition for E_1 to be minimal is satisfied by equating to zero the partial derivatives of the first-order of E_1 with respect to the unknown coefficients $p(m, n)$, $q(m, n)$ at each step i. Due to the form of the error function E_1 (equation (8)) a linear system of equations is obtained at each step i.

$$[S]^{(i)}\boldsymbol{x}^{(i)} = \boldsymbol{c}^{(i)}; \ \boldsymbol{x}^{(i)} = [p^{(i)}(0, 0), p^{(i)}(0, 1), \ldots, p^{(i)}(N, N), q^{(i)}(0, 1), \ldots, q^{(i)}(N, N)]^T \tag{9}$$

where the constant vector $\boldsymbol{c}$ depends on the normalization $q(0, 0) = 1$.

A slightly different approach[8] has been developed first by evaluating the partial derivatives of the error function $\varepsilon(7)$ and then substituting the nonlinear terms according to the filter estimated at the previous step. This procedure allows a fast convergence, if some requirements on phase specifications, on the starting denominator and on stability are satisfied. A wide discussion on this problem and a detailed procedure for two-dimensional recursive filters design can be found in ref. 3.

A different strategy consists in decoupling the amplitude response optimization from that of the phase response.

2.2. Amplitude response optimization

In this case a cascade of recursive filters like (2), optimal with respect to a desired amplitude characteristic, has to be determined. Good results can be obtained by using nonlinear optimization techniques[1,3] without any constraint on the phase response.

2.3. Phase response optimization

A nonlinear technique has been proposed[6] to optimize the group delay response of recursive filters, which gets very good results, when a linear phase behaviour is required, but it is unable to deal with other phase response specifications.

In the following a technique for designing phase filters is described according to the same procedure illustrated in equations (5)–(9). In this case only a phase function $\phi(\omega_1, \omega_2)$ is given, which means an input frequency characteristic:

$$F_k = \cos\phi_k + \mathrm{j}\sin\phi_k \tag{10}$$

The phase filter, as a cascade of sections like (3) has a complex frequency response at the fixed frequency pairs $(\omega_{1_k}, \omega_{2_k})$:

$$H_k = \exp\{-\mathrm{j}(N\omega_{1_k} + N\omega_{2_k} + 2\angle D(\mathrm{e}^{-\mathrm{j}\omega_{1_k}}, \mathrm{e}^{-\mathrm{j}\omega_{2_k}})\} \tag{11}$$

The coefficients $d(m, n)$ can be determined by minimizing the error function E_1 (equation (8)), for this specific case; this involves the solution of a linear system of equations (9) whose solving matrix can be written[9] at each step i as

$$S^{(i)}_{r,s} = \sum_{k=1}^{K} \{\cos[(m-u)\omega_{1_k} + (n-v)\omega_{2_k}]$$

$$-\cos\phi_k \cos[(N-m-u)\omega_{1_k} + (N-n-v)\omega_{2_k}]$$

$$+\sin\phi_k \sin[(N-m-u)\omega_{1_k} + (N-n-v)\omega_{2_k}]\}/|D^{(i-1)}(e^{-j\omega_{1_k}}, e^{-j\omega_{2_k}})|^2$$

$$\begin{cases} r = u(N+1)+v \\ s = m(N+1)+n \end{cases} \qquad \begin{cases} u, v = 0, 1, \ldots, N & u+v \neq 0 \\ m, n = 0, 1, \ldots, N & m+n \neq 0 \end{cases}$$

$N = 1$ or 2 is the chosen order for the two-dimensional all-pass phase filter. The constant term is given by

$$c^{(i)}_r = S^{(i)}_{r,0}$$

and the vector of unknown coefficients is:

$$\boldsymbol{x}^{(i)} = [d(0, 1), \ldots, d(N, N)]^T$$

The linear system can be easily modified when symmetric constraints are established. In this way any phase characteristic can be approximated with success provided it meets a realistic negative slope, so that the method is not confined to an all-pass phase equalizer design.

Table 1 shows a test for the algorithm. An all-pass phase filter has been simulated by using a stable denominator $d(m, n)$.[1] Its phase response has been used as input specification for the design algorithm without any initial condition, which means

$$|D^0(e^{-j\omega_{1k}}, e^{-j\omega_{2k}})|^2 = 1 \qquad \forall k$$

in equation (12). The evaluated coefficients are identical to those of the input filter with a residual error $E_1 \cong 3 \times 10^{-9}$ which means that, under ideal conditions, the method works very well.

We now check its performance in a more realistic situation.

TABLE 1.

	Input coefficients	Output coefficients
$d(0, 1)$	−1·78813	−1·7881298
$d(0, 2)$	0·8293	0·8292999
$d(1, 1)$	3·2064	3·2063999
$d(1, 2)$	−1·49271	−1·4927101
$d(2, 2)$	0·69823	0·6982297

3. Applications of phase filtering

3.1. Linear phase equalization

This is a classical problem always encountered whenever a two-dimensional recursive filter is used, unless for particularly simple situations with very weak constraints.[10]

Let a two-dimensional recursive filter like (2), having a certain nonlinear phase response, $\theta(\omega_1, \omega_2)$ be given. The input specifications for the algorithm are:

$$\phi(\omega_{1_k}, \omega_{2_k}) = -\tau_o(\omega_{1_k} + \omega_{2_k}) - \theta(\omega_{1_k}, \omega_{2_k})$$

As regards the minimization of (8) the constant delay τ_0 is a non-linear term and must be determined in an iterative way, which involves a very long procedure, when the initial conditions are very distant from the required minimum.

This problem has been considered[6] by using a second-order all-pass

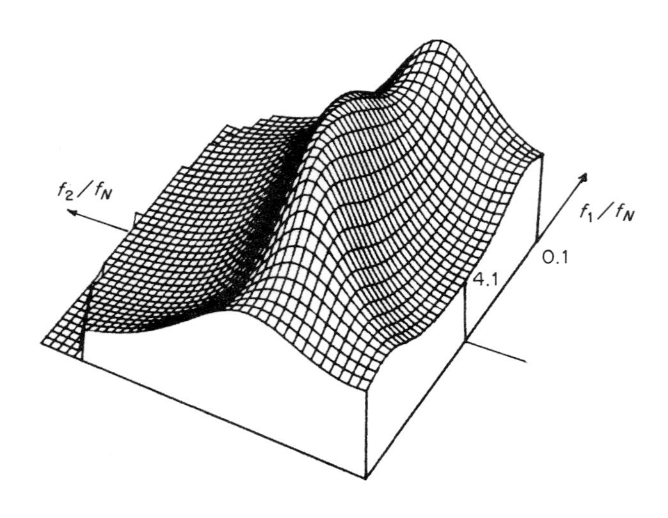

$$q(m,n) = \begin{bmatrix} 1 & -1{\cdot}78813 & 0{\cdot}82930 \\ -1{\cdot}78813 & 3{\cdot}20640 & -1{\cdot}49271 \\ 0{\cdot}82930 & -1{\cdot}49271 & 0{\cdot}69823 \end{bmatrix} \quad p(m,n) = \begin{bmatrix} 1 & -1{\cdot}62151 & 0{\cdot}99994 \\ -1{\cdot}62151 & 2{\cdot}63704 & -1{\cdot}62129 \\ 0{\cdot}99994 & -1{\cdot}62129 & 1{\cdot}00203 \end{bmatrix}$$

Fig. 1. Low-pass filter: (*a*) Group delay response; (*b*) filter coefficients. Cutoff frequency $f_c = 0{\cdot}1 f_N$.

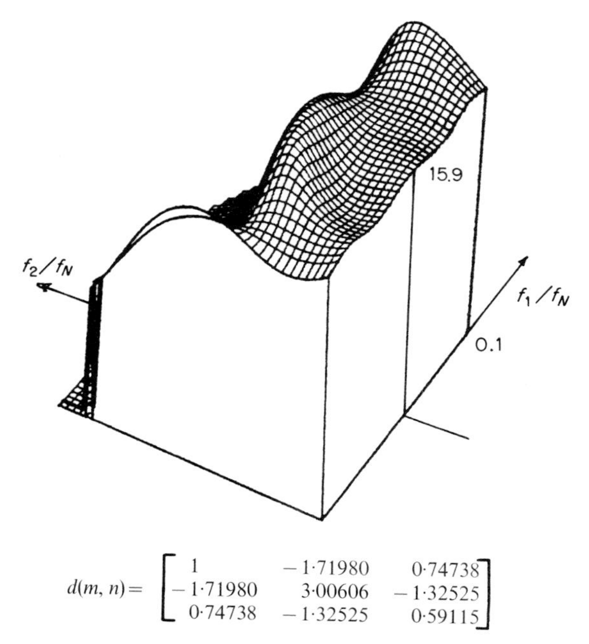

$$d(m, n) = \begin{bmatrix} 1 & -1{\cdot}71980 & 0{\cdot}74738 \\ -1{\cdot}71980 & 3{\cdot}00606 & -1{\cdot}32525 \\ 0{\cdot}74738 & -1{\cdot}32525 & 0{\cdot}59115 \end{bmatrix}$$

Fig. 2. Group delay reponse after equalization. All-pass phase filter coefficients are show below.

filter in order to equalize the group delay response of the filter in Fig. 1 which is an optimal low-pass filter [1] with cutoff frequency equal to one tenth of the Nyquist frequency.

Fig. 2(*a*) shows the group delay equalization by putting in cascade the low-pass filter with an all-pass equalizer whose coefficients are given in Fig. 2(*b*). The squared group delay variation about the mean, $\tau_0 = 19{\cdot}3558$, which turns out to be $J \approx 163$ against $J \approx 171$ in ref. 6.

3.2. Seismic signal processing

It has been shown[11] that, provided some approximations are satisfied, the propagation of acoustic waves, generated by impulse sources or reflectors, in a two-dimensional layered medium, can be modelled by a linear operator with frequency response:

$$F(\omega_x, \omega_t) = \begin{cases} \exp\left\{-\mathrm{j}\ h_0 \sqrt{\dfrac{\omega_t^2}{c^2} - \omega_x^2}\right\} & |\omega_t| > c|\omega_x| \\ 0 & \text{otherwise} \end{cases} \qquad (13)$$

where h_0 represents the depth of the source, c is the average acoustic velocity of the medium, (ω_x, ω_t) are the space and time frequencies respectively, and only upgoing waves have been considered.

The impulse response of such an operator, has dominant values in the space-time domain (x, t), along the hyperbola[11]

$$c^2t^2 - x^2 = h_0^2 \tag{14}$$

A half-plane fan-filter[3] is required, to remove the frequency content of the input signal in the band $|\omega_t| \leqq c|\omega_x|$, then a two-dimensional phase filter (13) with a half-plane impulse response (14) must be designed.

The problem was solved by using the technique described in the previous section where the filter structure was a rotated one-quadrant filter[9] with the constraint, $\Delta x = c\Delta t$ on the sampling intervals $\Delta x, \Delta t$. An all-pass phase filter has been designed with a phase specification like (13) and $h_0 = \Delta x$ and by using the same filter four times we get an impulse response as shown in Fig. 3.

In Fig. 4 an example of seismic signal processing is shown where an ideal diffraction pattern has been evaluated when a source at depth $h_0 = 20\Delta x$ explodes at time zero. The result of migrating such a response pattern with

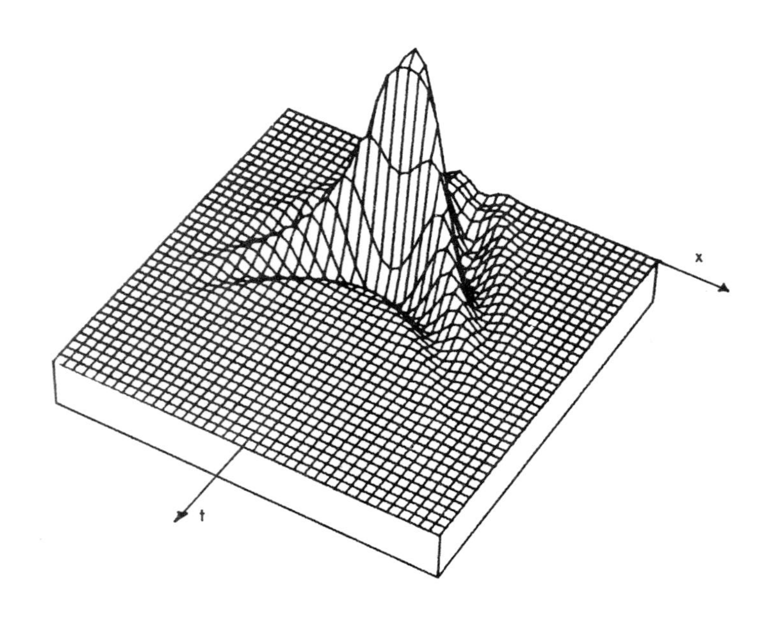

Fig. 3. Impulse response hyperbolic behaviour of the recursive downward continuation operator $(h_0 = 4\Delta x)$.

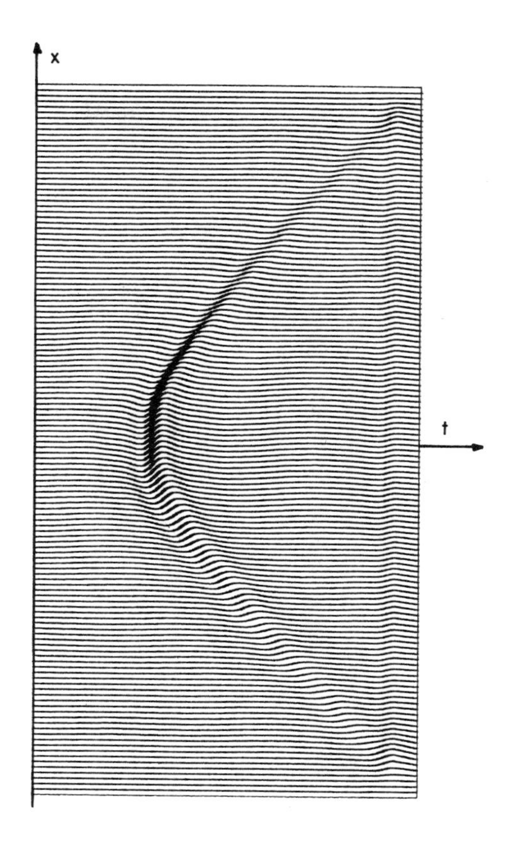

Fig. 4. Diffraction pattern: time section recorded when a single source emits an impulse (depth $h_0 = 20\Delta x$).

a two-dimensional recursive half-plane phase operator is shown in Fig. 5 by using 20 sections in cascade (8 coefficients per section).

3.3 Minimum phase filters

In this case a two-dimensional recursive filter like (2) has been designed according to a given amplitude characteristic with the minimum-phase constraint on both the numerator and denominator, which means that even the inverse filter is a realizable stable recursive operator. The amplitude responses of a low-pass filter and its correspondent high-pass inverse filter are shown in Fig. 6. Such a combination is very useful in pre-filtering an

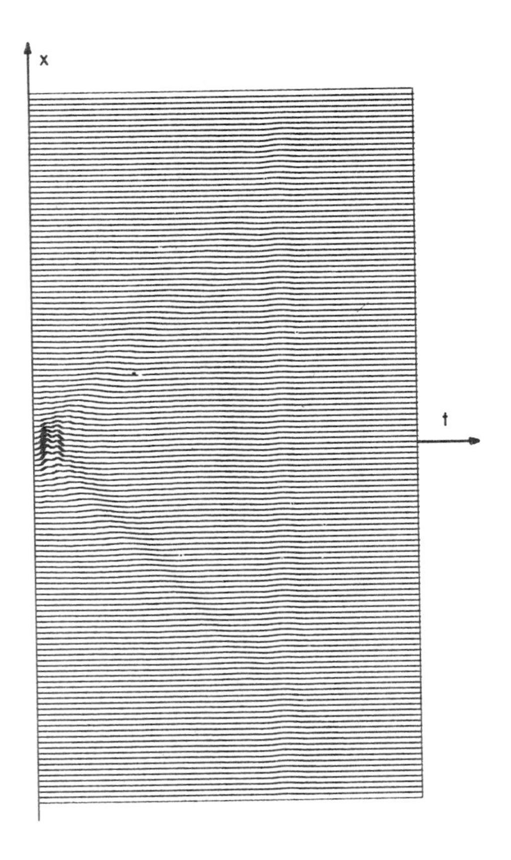

Fig. 5. Response pattern of Fig. 4 migrated with a cascade of 20 phase operators.

image for image data compression and subsequent amplitude response equalization. Moreover, due to the minimum phase constraint, the group delay response of these filters has small deviations about zero-phase behaviour.

4. Conclusions

Phase filtering can be an effective tool in two-dimensional signal processing. A design technique for any kind of phase operators and some of its more useful applications have been described in the paper. Further developments both in the design procedures and filter structure can lead to great improvements in two-dimensional recursive filter performance.

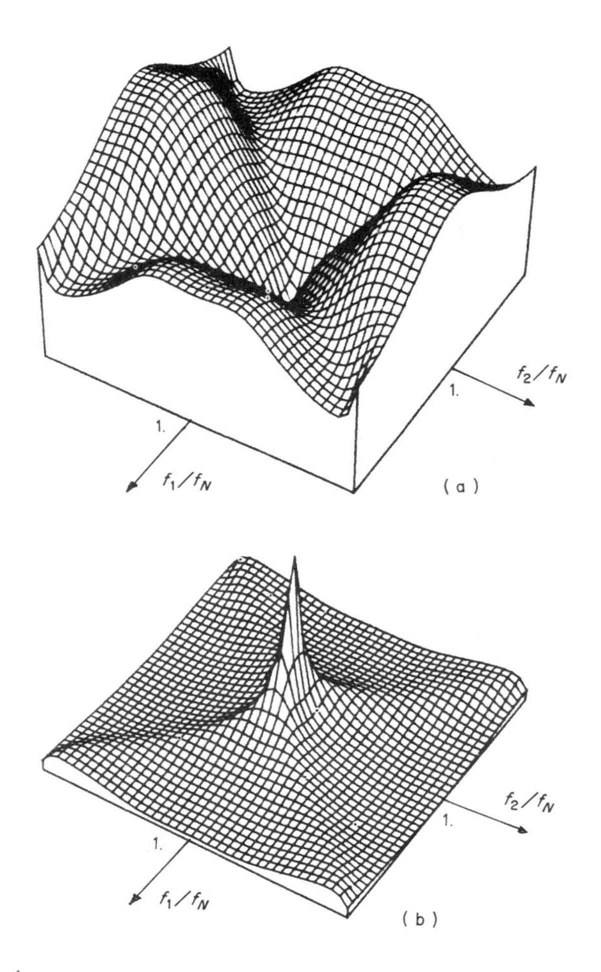

Fig. 6. Amplitude response of minimum-phase filters: (*a*) low-pass filter; (*b*) high-pass filter.

References

1. G. A. Maria and M. M. Fahmy, "An lp Design Technique for Two-Dimensional Digital Recursive Filters", *IEEE Trans. on Acoustics Speech and Signal Processing*, **ASSP-22**, pp. 15–21 (Feb. 1974).
2. M. P. Ekstrom and J. W. Woods, "Two-Dimensional Spectral Factorization with Applications in Recursive Digital Filtering", *IEEE Trans. on Acoustics Speech and Signal Processing*, **ASSP-24**, pp. 115–128 (April 1976).
3. G. Garibotto, "A New Approach To Half-Plane Recursive Filter Design", submitted for publication to *IEEE Trans. ASSP*.
4. R. M. Mersereau and D. E. Dudgeon, "Two-Dimensional Digital Filtering", *Proc. IEEE*, **63**, No. 4, pp. 610–623 (April 1975).

5. G. Garibotto, "On 2-D Recursive Digital Filter Design with Complex Frequency Specifications" *Proceedings of ICC '77, Chicago June* 1977, pp. 9B.2.202–206.
6. G. A. Maria and M. M. Fahmy, "lp Approximation of the Group Delay Response of One-and Two-Dimensional Filters", *IEEE Trans, on Circuits and Systems*, **CAS-21** No. 3, pp. 431–436 (May 1974).
7. C. K. Sanathanan and H. Tsukui, "Synthesis of Transfer Function from Frequency Response Data", *Int. J. System Science*, **5**, No. 4, pp. 41–54 (1974).
8. A. W. M. Enden, G. C. Groenendaal and E. Zee, "An Improved Complex-Curve Fitting Method", presented at CADEMICS, Hull, England, July 1977.
9. G. Garibotto, "2-D Recursive Phase Filters for the Solution of Two- Dimensional Wave Equations", *IEEE Trans. on Acoustics Speech and Signal Processing*, ASSP-27, No. 4, pp. 367–373 (Aug. 1979).
10. T. S. Huang, J. W. Burnett and A. G. Deczky, "The Importance of Phase in Image Processing Filters" *IEEE Trans. on Acoustics Speech and Signal Processing*, **ASSP-23**, No. 6 (Dec. 1975).
11. G. Bolondi, F. Rocca and S. Savelli, "A Frequency Domain Approach to Two-Dimensional Migration", 39th Meeting of the European Association of Exploration Geophysicists, Zagreb, Yugoslavia June 1977.

Part 2

TRANSFORMATIONS

Applications of Algebraic Numbers to Computation of Convolutions and DFTs with Few Multiplications

A. N. VENETSANOPOULOS

Department of Electrical Engineering, University of Toronto, Toronto, Canada

E. DUBOIS

INRS-Telecommunications, University of Quebec, Nun's Island, Canada

1. Introduction

In many special purpose digital signal processing applications, the cost of performing multiplications is significantly greater than that of additions. In such situations, algorithms which reduce the number of multiplications, perhaps with some increase in additions, may be attractive. Winograd[1] has shown that the number of multiplications required to perform certain polynomial multiplications related to computation of DFTs and convolutions depends on the underlying field of constants. Thus, by working in certain algebraic number fields, algorithms with a reduced number of multiplications may be devised, although a certain overhead may be incurred when transforming data into the new representation. In this paper, we describe the application of several rings and fields containing a cube root of unity to computation of DFTs and convolutions. In particular, we consider the applications to a radix-3 FFT which has no multiplications in the 3-point DFTs, and to number theoretic transforms.

2. Arithmetic with a cube root of unity

Let R be a commutative ring with identity and consider the extension ring

$$R(\theta)=\{a+b\theta:a,\ b\in R,\ \theta^2+\theta+1=0\}$$

θ is a cube root of unity, since $\theta^2=-\theta-1$ implies $\theta^3=-\theta^2-\theta=1$.

Operations in $R(\theta)$ are as follows:

$$(a+b\theta)+(c+d\theta)=(a+c)+(b+d)\theta$$

$$(a+b\theta)(c+d\theta)=(ac-bd)+[ad+b(c-d)]\theta$$

Multiplication requires three additions and four multiplications in R. It is possible to trade additions for multiplications by using the formula

$$(a+b\theta)(c+d\theta)=(ac-bd)+[(a+b)(c+d)-ac-2bd]\theta$$

which requires six additions and three multiplications in R. This may be useful if a multiplication takes longer than three additions. Also the addition $bd+bd=2bd$ my be implemented as a binary shift if desired.

3. Radix-3 FFT

The DFT of N points is given by

$$X(k)=\sum_{n=0}^{N-1} x(n)W^{nk}, \qquad k=0, \ldots, N-1 \tag{1}$$

where $W=\exp(-\mathrm{j}2\pi/N)$ and $X(k)$ and $x(n)$ are sequences of complex numbers. Assume that $N=3^M$. A decimation-in-time implementation of equation (1) for $N=27$ is shown in Fig. 1, using the notation described in ref. 2. The 3-point DFTs of Fig. 1 can be computed using four real multiplications and twelve real additions.[3]

If equation (1) is implemented in $\mathbb{R}(\theta)$, where $\mathbb{R}$ denotes the field of real numbers and

$$\theta=\exp(-\mathrm{j}2\pi/3)=-\frac{1}{2}-\frac{\sqrt{3}}{2}\mathrm{j}$$

then

$$W=\left[\cos\left(\frac{2\pi}{N}\right)+\sin\left(\frac{2\pi}{N}\right)/\sqrt{3}\right]+\left[2\ sin\left(\frac{2\pi}{N}\right)/\sqrt{3}\right]\theta$$

The 3-point transforms of Fig. 1 for this number system are shown in detail in Fig. 2. These require no multiplications, as shown explicitly by the following equations:

$$\begin{aligned}X(0)&=[x_1(0)+x_1(1)+x_1(2)]+[x_2(0)+x_2(1)+x_2(2)]\theta\\&=[x_1(0)-x_2(1)x_1(2)+x_2(2)]\\&\quad+[x_2(0)+x_1(1)-x_2(1)-x_1(2)]\theta\\X(2)&=[x_1(0)-x_1(1)+x_2(1)-x_2(2)]\\&\quad+[x_2(0)-x_1(1)+x_1(2)-x_2(2)]\theta\end{aligned} \tag{2}$$

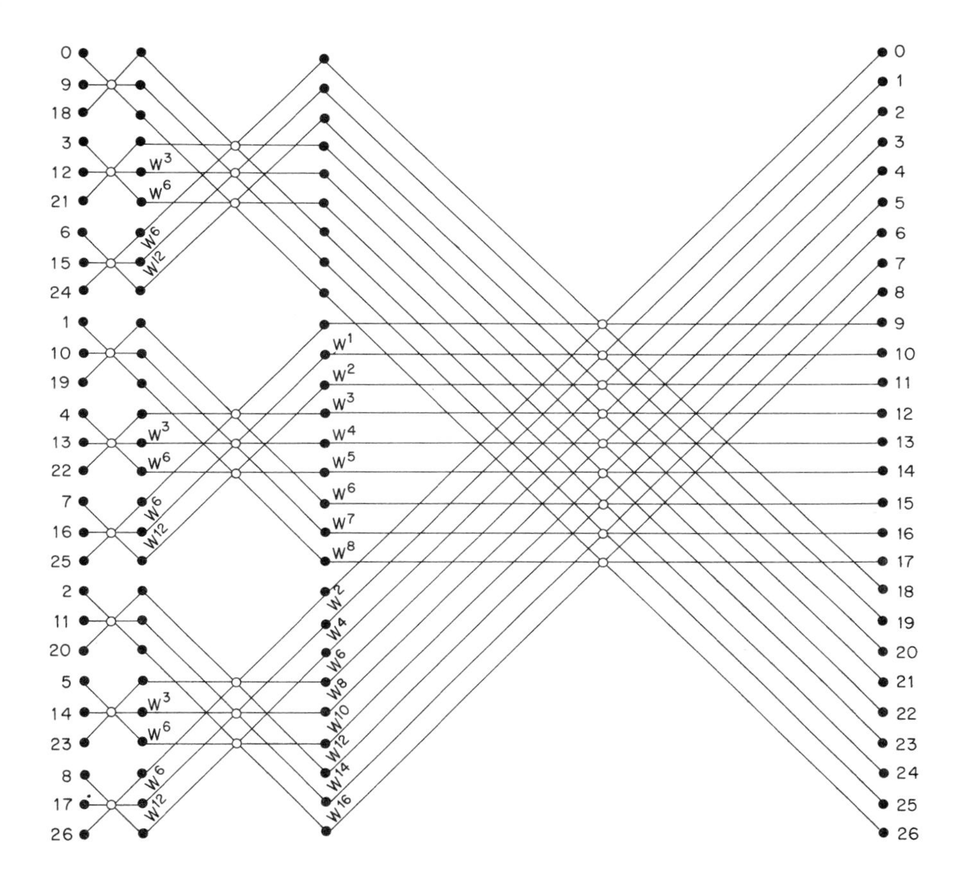

Fig. 1. A radix-3 in-place decimation-in-time FFT algorithm.

where $x(n) = x_1(n) + x_2(n)\theta$. Equation (2) can be evaluated with fourteen real additions. The only multiplications in the transform are those by the "twiddle factors" W^i which are shown in Fig. 1.

To obtain the DFT of a sequence of complex numbers using the above technique, the following equations must be used to transform input and output data between $\mathbb{C}$ and $\mathbb{R}(\theta)$.

$$a + bj \to \left(a - \frac{b}{\sqrt{3}}\right) - \frac{2b}{\sqrt{3}}\theta, \qquad a + b\theta \to \left(a - \frac{b}{2}\right) - \frac{\sqrt{3}b}{2}j$$

If either input or output are known to be real, the appropriate transformation can be waived. Although we have described the algorithm with

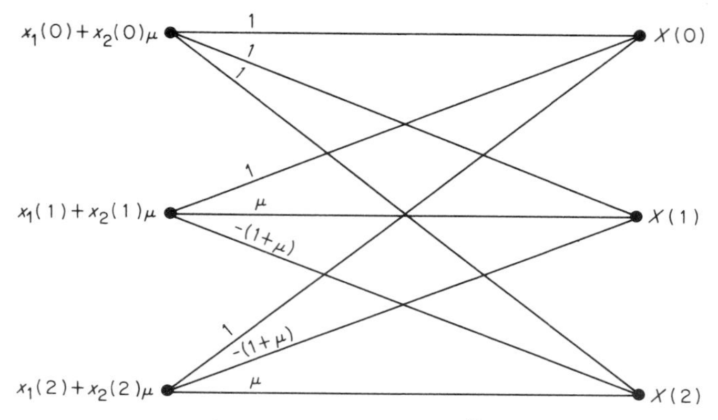

Fig. 2. Three-point DFT in $\mathbb{R}(\theta)$.

reference to the decimation-in-time implementation of Fig. 1, the technique applies to all the usual configurations a radix-3 FFT can take.

If the data are real, the above algorithm can be used to compute the DFT of sequences of length 2×3^M. This is accomplished by computing the transform of the length 3^M sequence $y(n) = x(2n) + x(2n+1)\theta$, and manipulating it in a fashion analogous to that for the complex-valued DFT.[4] If the transform is to be used to convolve real sequences, no data transformations are required. To convolve $x(n)$ and $h(n)$, the transforms $X(k)$ and $H(k)$ are computed in $\mathbb{R}(\theta)$, multiplied pointwise in $\mathbb{R}(\theta)$, and then inverse transformed. A technique to convolve length 2×3^M real sequences is given in ref. 5.

In an $N = 3^M$ point radix-3 FFT, there are $(\frac{2}{3}M - 1)N + 1$ non-trivial complex multiplications by twiddle factors and $MN/3$ three point DFTs. Use of the above algorithm eliminates the $4MN/3$ real multiplications associated with these 3-point DFTs, while leaving the number of real multiplications associated with the twiddle factors fixed. In the limit for large N, the number of real multiplications is reduced by a factor of $33\frac{1}{3}\%$, if multiplication algorithms in $\mathbb{C}$ and $\mathbb{R}(\theta)$ with four real multiplications are used, and by a factor of 40%, if multiplication algorithms in $\mathbb{C}$ and $\mathbb{R}(\theta)$ with three real multiplications are used. However the number of additions is increased and the total computation savings depend on the relative costs of addition and multiplication on the processor being used. If the cost of a multiplication is r times that of an addition, then for large N the relative cost of the arithmetic operations in the new algorithm to those in the standard algorithm is given by

$$\frac{2r+5}{3r+4}$$

if multiplication algorithms in $\mathbb{C}$ and $\mathbb{R}(\theta)$ with four real multiplications are used. This ratio is shown in Fig. 3 as a function of r; it is unity for $r=1$ and decreases monotonically to $\frac{2}{3}$. If the multiplication algorithms in $\mathbb{C}$ and $\mathbb{R}(\theta)$ with three real multiplications are used, the corresponding ratio is

$$\frac{3r+13}{5r+11}$$

which is unity for $r=1$ and decreases monotonically to $\frac{3}{5}$.

Similar comparisons can be made with radix-2 and radix-4 FFTs. Restricting the situation to the case where multiplication in $\mathbb{C}$ and $\mathbb{R}(\theta)$ requires four real multiplications, for large N the relative cost of the arithmetic operations in the new algorithm and in a radix-2 algorithm is given by

$$\log_3 2 \frac{8r+20}{6r+9} = \frac{5{\cdot}05r+12{\cdot}62}{6r+9}$$

This ratio is 1·18 for $r=1$, becomes equal to unity for $r=3{\cdot}81$ and decreases monotonically to 0·84. For radix-4, the ratio is

$$\log_3 4 \frac{16r+40}{18r+33} = \frac{20{\cdot}19r+50{\cdot}47}{18r+33}$$

which is 1·39 at $r=1$ and decreases monotonically to 1·12. These two ratios are also shown in Fig. 3. The new algorithm can thus be more efficient than radix-2 but is less efficient than radix-4. It must be understood that these ratios represent relative efficiencies, as a radix-2 and radix-3 algorithm cannot both exist for the same N.

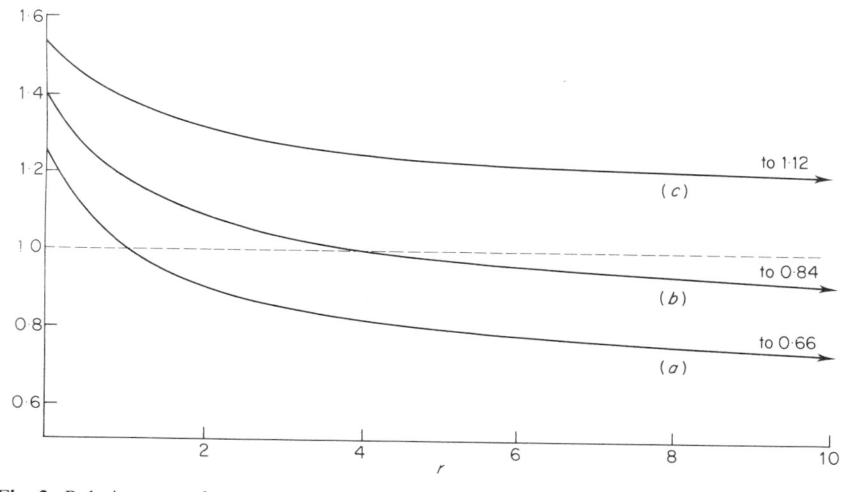

Fig. 3. Relative cost of computation between new radix-3 algorithm and (*a*) standard radix-3, (*b*) radix-2, (*c*) radix-4 FFT algorithms.

To extend this technique to higher radices, new bases $\{(\theta_1, \theta_2)\}$ are needed, for the complex field. The properties desired are: (a) that arithmetic in these bases has simplicity comparable to standard complex arithmetic and (b) that a p-point DFT has a reduced number of multiplications for some p. To date the authors have not discovered any further examples satisfying these two properties. In particular, the basis $(1, \psi)$ where ψ is a primitive pth root of unity does not satisfy them for $p > 3$.

4. Application to number theoretic transforms

A number theoretic transform is a transform defined in a ring R of residues of algebraic integers having the DFT structure, which maps cyclic convolution isomorphically into pointwise product:[6,7]

$$U\colon X(k) = \sum_{n=0}^{N-1} x(n)\alpha^{kn}, \qquad k = 0, \ldots, N-1$$

and

$$U(x * y) = Ux \,.\, Uy$$

where $x * y$ denotes cyclic convolution. A necessary condition for these properties to hold is that α be a primitive Nth root of unity in R.

Generally, the ring R is chosen to be a ring of integers, or complex integers, modulo an integer M. By appropriate choice of M and α, transforms which require no multiplications can be constructed. However, these suffer from the disadvantage that the wordlength required is proportional to the sequence length. Thus techniques which can give greater sequence lengths for given wordlength without introducing multiplications are of interest.

The most well known moduli used with number theoretic transforms are Fermat numbers: $F_t = 2^{2^t} + 1$. Using $\alpha = 2$, transform lengths of 2^{t+1} can be achieved and with $\alpha = \sqrt{2} \triangleq 2^{t-2}(2^{t-1} - 1)$ lengths 2^{t+2} are obtained with transforms in Z_{F_t}. By working in the ring $Z_{F_t}(\theta)$, these transform lengths can be increased by a factor of 3,[7] using $\alpha = 2\theta$ or $\alpha = \sqrt{2\theta}$ (other values may be more advantageous). Then, for real data, the sequence lengths which can be convolved can be further doubled using techniques described in ref. 8. Arithmetic is carried out as indicated in Section 2.

For illustration, consider $M = 2^8 + 1$ and $N = 3 \times 2^4$, with $\alpha = 2^{11}\theta$. (2^{11} has order 16, and has the property $(2^{11})^3 = 2^{33} = 2$.) The transform can be decomposed as three 16-point transforms with $\alpha = (2^{11}\theta)^3 = 2$, followed by sixteen 3-point transforms with $\alpha = (2^{11}\theta)^{16} = \theta$. Since 3 and 16 are relatively prime, the prime factor algorithm can be used, obviating the need for intermediate multiplications by twiddle factors.

The 3-point transforms can be implemented exactly as in Fig. 2, requiring fourteen additions, as given by equation (2). The flowgraph for the entire transform is shown in Fig. 4, where the 16-point transforms can be implemented using any standard radix-2 algorithm.

Useful results can also be obtained by taking R to be the ring of Gaussian integers modulo M, $Z_M[i]$. In this case, transforms in $Z_M[i](\theta)$ can be used to convolve sequences of complex numbers, giving an increase in sequence length by a factor of six over comparable algorithms in $Z_M[i]$. Thus for example, if a transform of length N having no multiplications exists in $Z_M[i]$ with $\alpha = 1 + i$, then one of length $3N$ exists in $Z_M[i](\theta)$ with $\alpha =$

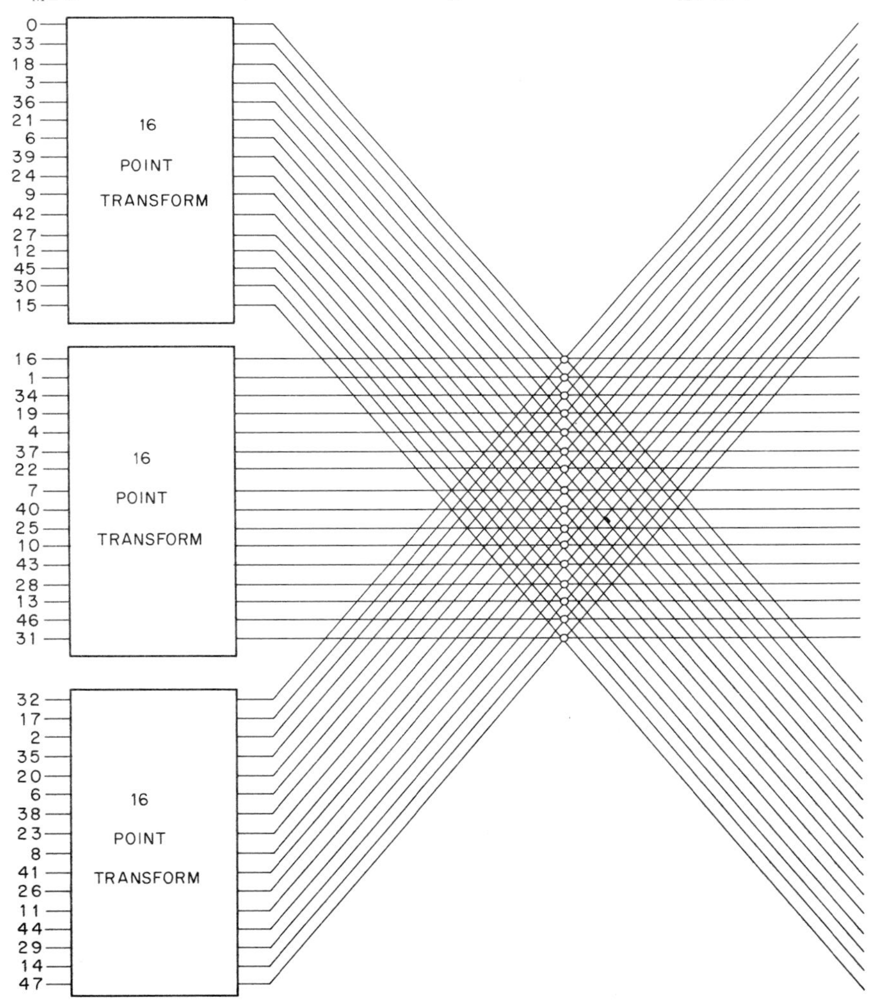

Fig. 4. Forty-eight-point DFT in $Z[\theta](F_t)$.

$(1+i)\theta$, and the techniques of ref. 8 can be used to convolve sequences of length $6N$. These transforms are most useful with the pseudo-Mersenne and pseudo-Fermat transforms.[9,10]

5. Conclusion

Techniques for computation of convolutions and DFTs using arithmetic in rings and fields containing a cube root of unity have been described. An algorithm for computing the radix-3 FFT using arithmetic in $\mathbb{R}(\theta)$ uses $33\frac{1}{3}\%$ or 40% fewer multiplications than the conventional radix-3 algorithm. Also, number theoretic transforms exhibiting increased sequence lengths can be obtained using rings of modular integers containing a cube root of unity. These results show that use of number systems other than the real and complex fields and their integers can lead to computational savings in digital signal processing applications.

Acknowledgment

This work was supported in part by the National Research Council of Canada under Grant No. A-7397.

References

1. S. Winograd, "The effect of the field of constants on the number of multiplications", *Proc. 16th Annual Symposium on Foundations of Computer Science*, pp. 1–2 (1975).
2. L. R. Rabiner and B. Gold, "Theory and Application of Digital Signal Processing", pp. 435–437. Englewood Cliffs, N.J.: Prentice-Hall, 1975.
3. S. Winograd, "On computing the discrete Fourier transform", *Proc. of the National Academy of Sciences, USA*, **73**, pp. 1005–1006 (April 1976).
4. J. W. Cooley, P. A. W. Lewis, and P. D. Welch, "The fast Fourier transform algorithm: Programming considerations in the calculation of sine, cosine and Laplace transforms", *J. Sound Vib.*, **12**, pp. 315–337 (July 1970).
5. E. Dubois and A. N. Venetsanopoulos, "A new algorithm for the radix-3 FFT", *IEEE Trans. on Acoustics, Speech and Signal Processing*, **ASSP-26** (June 1978).
6. R. C. Agarwal and C. S. Burrus, "Number theoretic transforms to implement fast digital convolution", *Proc. IEEE*, **63**, pp. 550–560 (April 1975).
7. E. Dubois and A. N. Venetsanopoulos, "The discrete Fourier transform over finite rings with application to fast convolution", *IEEE Trans. on Computers*, **C-27** (July 1978).
8. E. Dubois and A. N. Venetsanopoulos, "Convolution using a conjugate symmetry property for the generalized discrete Fourier transform", *IEEE Trans. on Acoustics, Speech, and Signal Processing*, **ASSP-26**, pp. 165–170 (April 1978).
9. H. J. Nussbaumer, "Digital filtering using complex Mersenne transforms", *IBM J. Res. Dev.*, **20**, pp. 498–504 (Sept. 1976).
10. H. J. Nussbaumer, "Digital filtering using pseudo Fermat number transforms", *IEEE Trans. on Acoustics, Speech, and Signal Processing*, **ASSP-26**, pp. 79–83 (Feb. 1977).

Shifted Discrete Fourier Transforms

L. P. JAROSLAVSKI

Institute of Information Transmission Problems Moscow, U.S.S.R.

The notion is introduced of "shifted discrete Fourier transforms" which are a generalization of the well known discrete Fourier transform. Properties of shifted discrete Fourier transforms and their advantages over discrete Fourier transforms are summarized, and possible applications are indicated.

The discrete Fourier transform (DFT) defined for a sequence $\{a_k\}$ ($k=0, 1, \ldots, N-1$), as

$$\alpha_r = \frac{1}{\sqrt{N}} \sum_{k=0}^{N-1} a_k \exp\left(i2\pi \frac{kr}{N}\right) \tag{1}$$

plays a leading role in digital signal processing and digital holography because one may regard it as a discrete representation of the continuous Fourier transform. Usually, the DFT is derived from the continuous Fourier transform by applying the sampling theorem to signals and their Fourier spectra (see, for instance, ref. 1). In doing so, one usually disregards the fact that the sample raster of the signal and its spectrum may be shifted with respect to coordinate systems where the continuous signal and its spectrum are considered. If the zero sample of signal $a(t)$ with respect to the origin of the signal coordinates is $u\Delta t$ (where Δt is signal sampling interval) and if the zero sample of signal spectrum $\alpha(s)$ with respect to the origin of spectrum coordinates is $v\Delta s$ (where Δs is spectrum sampling interval), direct and inverse Fourier transforms may be represented in the discrete form as follows:

$$\alpha_r^{u,v} = \frac{1}{\sqrt{N}} \sum_{k=0}^{N-1} a_k \exp\left[i2\pi \frac{(k+u)(r+v)}{N}\right] \tag{2a}$$

$$a_k^{u,v} = \frac{1}{\sqrt{N}} \sum_{r=0}^{N-1} \alpha_r^{u,v} \exp\left[-i2\pi \frac{(k+u)(r+v)}{N}\right]. \tag{2b}$$

Superscripts u and v of a_k in (2b) stress the fact that $a_k^{u,v}$ is defined for any k

as distinct from the original sequence $\{a_k\}$ of the signal sample $a(t)$ defined for $k=0, 1, \ldots, N-1$. For $k=0, 1, \ldots, N-1$, $\{a_k^{u,v}\}$ coincides with the original sequence $\{a_k\}$.

The pair of transforms (2*a*, 2*b*) will be referred to as direct and inverse shifted discrete Fourier transforms and denoted as SDFT(u, v) and RSDFT(u, v), respectively. The standard discrete Fourier transform (equation (1)) is, obviously, SDFT(0, 0).

SDFT(u, v) are expressed in terms of DFT as follows:

$$\alpha_r^{(u,v)} = \left(\mathrm{DFT}\left\{a_k \exp\left[i2\pi \frac{v(k+u)}{N}\right]\right\}\right) \exp\left(i2\pi \frac{ur}{N}\right). \tag{3}$$

For integer u and v, SDFT(u, v) boils down to a cyclic shift of DFT of cyclically shifted sequences, and, therefore, properties of SDFT(u, v) coincide with those of the DFT for integer (u, v). This is not the case, however, for fractional (u, v). The basic properties of SDFT are given in Table 1.

As one may see from the comparison of these properties, the SDFT (u, v) features some peculiarities. It has a more general law of sequence continuation over numbers k and r different from $0, 1, \ldots, N-1$) (row 2 of the table), a more general definition of even and odd sequences (row 7 of the table), a more general formula of signal restoration through its spectrum enabling interpolation (determination of intermediate samples) of signal by means of the pair SDFT(u, v) and inverse SDFT(p, q) with appropriately selected (u, v) and (p, q) (see rows 11 *a*, *b* of the table), and a more general convolution theorem. (rows 12 *a*, *b* of the table). For interpolation it should be noted, that if p in the inverse SDFT(p, q) is made dependent on k, one may determine arbitrarily situated intermediate samples of $\{a_k\}$; this is important for signal coordinate transformations. This is valid also for determination of the intermediate samples of the convolution of signals by means of the SDFT.

Due to these peculiarities, some forms of SDFT prove to be more useful in digital signal processing than DFT. For instance, it may be shown that SDFT $(\frac{1}{2}, 0)$ is useful in signal convolution computations for even continuation of signals (in order to prevent boundary effects occuring at cyclic convolution). Since an even symmetry signal has SDFT$(\frac{1}{2}, 0)$ of odd symmetry (see row 7 of the table), a simple combined algorithm is possible for the SDFT$(\frac{1}{2}, 0)$ of two signals simultaneously. Thus the number of operations required for SDFT$(\frac{1}{2}, 0)$ of one signal does not increase in spite of duplication of sequence lengths under even continuation. This advantage is essential for two-dimensional signals, e.g. for processing of pictures by means of the two-dimensional SDFT.

Some versions of SDFT were suggested in the literature as an alternative to DFT, but no connection between them was demonstrated. For example,

TABLE 1. SDFT properties

	Signal	SDFT
(1)	$a_k^{u,v} = \frac{1}{\sqrt{N}} \sum_{r=0}^{N-1} \alpha_r^{u,v}$ $\times \exp\left[-i2\pi \frac{(k+u)(r+v)}{N}\right].$	$\alpha_r^{u,v} = \frac{1}{\sqrt{N}} \sum_{k=0}^{N-1} a_k$ $\exp\left[i2\pi \frac{(k+u)(r+v)}{N}\right]$
(2)	$a_{k+hN}^{u,v} = a_k \exp(-i2\pi hv)$	$\alpha_{r+gN}^{u,v} = \alpha_N^{u,v} \exp(i2\pi gv)$
(3)	$a_k^* \exp -i2\pi \frac{2v(k+u)}{N}$ a_k^*	$(\alpha_{N-r}^{u,v})^* \exp[i2\pi u]$. If $2v$ is integer, then $(\alpha_{N-2v-r})^* \exp i2\pi u$
(4)	$(a_{N-k}^{u,v})^* \exp(-i2\pi v)$: If $2u$ is integer, then $(a_{N-2u-k})^* \exp(-i2\pi v)$	$\alpha_r^* \exp\left[i2\pi \frac{2u(r+v)}{N}\right]$ α_r^*
(5)	$a_{N-k-1}^{u,v} \exp\left[-i2\pi \frac{(k+u)(2v-1)}{N}\right]$: If $(2u)$ and $(2v)$ are integers, then $a_{N-2u-k} \cdot (-1)^{2u}$	$\alpha_{N-r-1}^{u,v} \exp\left[i2\pi \frac{(r+v)(2u-1)}{N}\right]$ $\exp\left\{-i2\pi\left[(u-v) + \frac{(2u-1)(2v-1)}{N}\right]\right.$ $(-1)^{2v} \alpha_{N-2v-r}$
(6)	$\alpha_k = \pm a_k^* \exp\left[-i2\pi \frac{2v(k+u)}{N}\right]$ $a_k = \pm a_k^*$	$\alpha_r^{u,v} = \pm(\alpha_{N-r}^{u,v})^* \exp i2\pi u$ If $2v$ is integer, then $\alpha_r^{u,v} = \pm(\alpha_{N-2v-r}^{u,v})^* \exp(i2\pi u)$
	$2u$, $2v$ are integers	
(7)	$a_k^{u,v} = \pm a_{N-2u-k}^{u,v}$	$\alpha_r^{u,v} = \pm \alpha_{N-2v-r}^{u,v} (-1)^{2(u+v)}$
	$2u$, $2v$ are integers	
(8)	$a_k^{u,v} = (a_k^{u,v})^* = \pm a_{N-2u-k}^{u,v}$	$\alpha_r^{u,v} = (\alpha_{N-2v-r}^{u,v})^* \exp(i2\pi u)$ $= \pm \alpha_{N-2v-r}^{u,v} \exp[-i2\pi(u+v)]$
(9)	$a_{k+k_0}^{u,v}$	$\alpha_r^{u,v} \exp\left[-i2\pi \frac{k_0(r+v)}{N}\right]$
(10)	$a_k^{u,v} \exp\left[i2\pi \frac{r_0(k+u)}{N}\right]$	$a_{r+r_0}^{u,v}$

Table **1** *(Continued)*

Signal	SDFT
(11*a*) $^{r_0}a^{p,q/u,v} = \frac{1}{\sqrt{N}} \sum_{r=0}^{N-1} \alpha_{r+r_0}^{u,v} \exp\left[-i2\pi \frac{(k+p)(r+q)}{N}\right]$ $= \sum_{n=0}^{N-1} \left(a_n \exp\left[i2\pi \frac{(n+u)(r_0+v-q)}{N}\right]\right) \frac{\sin \pi(k-n+p-u)}{N \sin[\pi(k-n+p-u)/N]}$ $\exp\left[-i2\pi(k-n+p-u)\left(q+\frac{N-1}{2}\right)/N\right]$	
For $r_0+v=q=-(N-1)/2$ (11*b*) $^{r_0}\alpha^{p,q/u,v} = \sum_{n=0}^{N-1} \alpha_n \frac{\sin \pi(n-k+u-p)}{\sin[\pi(n-k+u-p)/N]}$	
$r_a, r_b c_k^{u_c,v_c} = \frac{1}{\sqrt{N}} \sum_{r=0}^{N-1} \alpha_{r+r_a}^{u_a,v_a} \beta_{r+rb}^{u_b,v_b} \exp\left[-i2\pi \frac{(k+u_c)(r+v_c)}{N}\right]$ $= \frac{1}{\sqrt{N}} \sum_{n=0}^{N-1} a_n \exp\left[i2\pi \frac{(n+u_a)(r_a+v_a-v_c)}{N}\right] b_{k-n+[u_c-(u_a+u_b)]}^{\mathrm{int}}$ (12*a*) $b_k^{\mathrm{int}} = \sum_{m=0}^{N-1} (b_m^{u_b,v_b} \exp[i2\pi(m+u_b)(r_b+v_b-v_c)/N]).$ $\frac{\sin \pi(m-k)}{N \sin[\pi(m-k)/N]} \exp\left[i2\pi(m-k)\left(v_c+\frac{N-1}{2}\right)/N\right]$	
If $[u_c-(u_a+u_b)]$ is integer: (12*b*) $^{r_a,r_b}c_k^{u_c,v_c} = \frac{1}{\sqrt{N}} \sum_{n=0}^{N-1} (a_n \exp[i2\pi(n+u_a)(r_a+v_a-v_c)/N])$ $b_{k-n+[u_c-(u_a+u_b)]}^{u_b,v_b} \exp[i2\pi(k-n+u_c-u_a)(r_b+v_b-v_c)/N]$	

SDFT(0, $\frac{1}{2}$) was considered in ref. 2 as a more convenient transformation of real-valued sequences; Ahmed *et al.*[3,4] introduced a so-called discrete cosine transform, which is the SDFT($\frac{1}{2}$, 0) of even continued sequences. Owing to the fact that even-continued (in the sense of row 7 of the table) signals correspond to this transform, it is very useful for signal coding, and in particular, for picture coding. In ref. 5 the so called sine transform is introduced, which is in fact the SDFT(1, 1) of odd-continued sequences of $2N$ samples with zero-valued $(N-1)$th and $(2N-1)$th samples.

To conclude let us note that from the computational point of view it is useful to modify the definitions of equation (2) of the SDFT by removing the constant coefficients $\exp(\pm i2\pi(uv/N))$. This will give us the following pair of transforms:

$$\alpha_r^{u,v} = \frac{1}{\sqrt{N}} \sum_{k=0}^{N-1} a_k \exp\left(i2\pi \frac{kv}{N}\right) \exp\left(i2\pi \frac{k+u}{N} r\right) \tag{4a}$$

$$a_k^{u,v} = \frac{1}{\sqrt{N}} \sum_{k=0}^{N-1} \alpha_r^{u,v} \exp\left(-i2\pi \frac{ru}{N}\right) \exp\left(-i2\pi \frac{(r+v)}{N} k\right) \tag{4b}$$

with practically the same properties as the SDFT, but more convenient for computation.

References

1. B. Gold and Ch. M. Rader, "Digital Processing of Signals." New York: McGraw-Hill, 1969.
2. J. L. Vernet, "Real signals fast Fourier transform. Storage capacity and step number reduction by means of an odd discrete Fourier transform," *Proc. IEEE*, **59**, No. 10 (1971).
3. N. Ahmed, N. Natarajan, and K. R. Rao, "Discrete Cosine Transform," *IEEE Trans. on Computers*, **C-23**, No. 1 (Jan. 1974).
4. N. Ahmed and K. R. Rao, "Orthogonal Transforms for Digital Signal Processing," Berlin: Springer-Verlag, 1975.
5. Jain, A. K. and Angel E. "Image restoration, modelling and reduction of Dimensionality", *IEEE Trans. on Computers*, **C-23**, No. 5 (May 1974).

Fast Computation of Toeplitz Forms under Narrow Band Conditions with Applications to Spectral Estimation

U. STEIMEL

Institut für Informatik der Universität Bonn, Bonn, German Federal Republic

1. Introduction

This paper deals with the problem of rapidly computing Toeplitz forms in finite sequences of observations on stochastic processes, which are real, zero mean, stationary, and narrow band. As an application we mention the discrete Gauss–Gauss detection problem with unknown signal-to-noise ratio,[1] where the detector sometimes decides with support of a test function which is a Toeplitz form. The idea of this paper is to use efficient algorithms which are correct for circulant forms and finite narrow band sequences, for a fast approximate computation of Toeplitz forms. These algorithms were developed by Böhme.[2] We also treat the case that the finite sequences are approximately band-limited. All investigations about resulting errors are asymptotic, so that the algorithms may be applied in practice when the number of observations is large.

A Toeplitz form is approximated by a suitable circulant form. Considering the approximation error as a function of the number T of observations, we give in Section 2 conditions such that the normalized bias error and the normalized standard error asymptotically vanish as T tends to infinity. A special case shows when the approximation yields a consistent estimate.

In Section 3, we describe some known algorithms for a fast computation of circulant forms in finite sequences which are exactly narrow band. An exact band limitation of the process generally does not imply the same to finite observation sequences. Therefore we give conditions under which the algorithms compute consistent estimates of the circulant forms, if the finite sequences are approximately band limited.

In the final section, we consider estimates of the correlation functions and the spectral densities at single points which are Toeplitz forms as used, for example, by Anderson[3] or Hannan.[4] By using the algorithms for circulant forms we get fast computable estimators for the correlation functions and spectral densities, respectively. Finally, we discuss the computation time for our estimators.

2. Approximation of a Toeplitz form by a circulant form

Let $x=(x(0), \ldots, x(T-1))$ be a finite sequence of T observations on a real, zero mean, stationary process $\{x(t)\}$. Actually, we use only weak stationarity up to the fourth moment in this paper. A Toeplitz form in x is

$$q_g = \sum_{t,r=0}^{T-1} g(t-r)x(t)x(r)$$

where we may assume that $g(t)=g(-t)$. Defining

$$h(t)=h(-t)=g(t)-t/T(g(t)-g(T-t)) \qquad t=0, \ldots, T-1$$

following Pearl[5] and substituting h for g in the Toeplitz form one gets a circulant form q_h which can be used as a reasonable estimator for q_g. To show this we use the dependence of q_h and q_g on T and investigate the asymptotic behaviour of the normalized bias error

$$e_{\mathrm{b}} = \mathrm{E}(q_g - q_h)/\mathrm{E}q_g$$

and the normalized standard error

$$e_{\mathrm{s}} = \sqrt{[\mathrm{Var}(q_g - q_h)]}/\mathrm{E}q_g$$

where E and Var denote expectation and variance, respectively. If

$$R(s) = \mathrm{E}x(t)x(t+s)$$

is the correlation function of $\{x(t)\}$, we have two statements about e_{b}:

(P1) $\lim_{T\to\infty} e_{\mathrm{b}}=0$, if $\lim_{T\to\infty} \sum_{s=1}^{T-1} s|g(s)|/\mathrm{E}q_g=0$.

(P2) $\lim_{T\to\infty} e_{\mathrm{b}}=0$, if there exist numbers b, $c>0$, such that $s^b R(s)$ is absolutely summable and $|g(s)/\mathrm{E}q_g| \leqq c/T$ for $0 \leqq s < T$ and all T.

Both statements immediately follow by use of

$$q_g - q_h = \sum_{s=-(T-1)}^{T-1} |s|(g(s) - g(T-|s|))c(s) \tag{1}$$

where

$$c(s)=c(-s)=\frac{1}{T}\sum_{t=0}^{T-1-s} x(t)x(t+s) \quad \text{for } s=0, \ldots, T-1$$

The variance of $q_g - q_h$ involves moments of fourth order. If $\{x(t)\}$ is a Gauss-process, we know that

$$\mathrm{E}x(t)x(r)x(t')x(r')=R(t-r)R(t'-r')+R(t-t')R(r-r')+R(t-r')R(r-t')$$

whereas in the non-Gaussian case we have to add another term called the fourth-order cumulant $\kappa(t-r, t-t', t-r')$. Hence, we can state two propositions about the normalized standard error:

(P3) $\lim_{T\to\infty} e_s = 0$, if $\lim_{T\to\infty} \sum_{s=1}^{T-1} s|g(s)|/\mathrm{E}q_g = 0$
and
$\max_{-T<r,s,t<T} |\kappa(r, s, t)|$ is uniformly bounded for all T.

(P4) $\lim_{T\to\infty} e_s = 0$, if the following assumptions are satisfied:
(i) $R(s)$ is absolutely summable;
(ii) there exist a number $c>0$ and a sequence $\{K_T\}$ of integers depending on T such that $K_T\to\infty$ and $K_T/T\to 0$ as $T\to\infty$ and $g(s)=0$ for $K_T<s<T$ and $|g(s)/\mathrm{E}q_g| \leqq c/T$ for $0\leqq s \leqq K_T$;
(iii) $\sum_{r,s,t=-\infty}^{\infty} |\kappa(r, s, t)|$ is finite.

Proposition (P3) is a direct consequence of

$$\mathrm{Var}(q_g - q_h) = \sum_{r,s=-(T-1)}^{T-1} |rs|[g(r)-g(T-|r|)] \times [g(s)-g(T-|s|)]\mathrm{Cov}(c(r), c(s)) \tag{2}$$

where

$$\mathrm{Cov}(c(r), c(s)) = \frac{1}{T^2}\sum_{t=0}^{T-|r|-1}\sum_{t'=0}^{T-|s|-1}[R(t-t')R(t-t'+r-s) + R(t-t'-s)R(t-t'+r) + \kappa(r, t'-t, t'-t+s)] \tag{3}$$

Proposition (P4) can be proved as follows. Equation (2) and the assumptions imply

$$e_s^2 \leqq c^2\left(\sum_{\substack{r,s=-K_T\\ \neq 0}}^{K_T} + 2\sum_{|r|=T-K_T}^{T-1}\sum_{\substack{s=-K_T\\ s\neq 0}}^{K_T} + \sum_{|r|,|s|=T-K_T}^{T-1}\right)|\mathrm{Cov}(c(r), c(s))|$$

Using equation (3) the first sum except for the non-Gaussian part is bounded by

$$c^2 \sum_{\substack{r,s=-K_T \\ \neq 0}}^{K_T} |\text{Cov}(c(r), c(s))|$$

$$\leqq \frac{c^2}{T}\left(\sum_{r=-(T-1)}^{T-1} \sum_{s=r-2K_T}^{r+2K_T} + \sum_{r,s=-(T+K_T-1)}^{T+K_T-1} \right) 2K_T |R(r)R(s)|$$

$$\leqq \frac{K_T}{T} 4c^2 \left(\sum_{s=-\infty}^{\infty} |R(s)| \right)^2$$

The non-Gaussian part of the first sum is obviously smaller than $(c^2/T)\Sigma_{r,s,t=-\infty}^{\infty} |\kappa(r, s, t)|$. Handling the other sums similarly, we have

$$e_s^2 \leqq \frac{K_T}{T} 20c^2 \sum_{s=-\infty}^{\infty} |R(s)|^2 + \frac{1}{T} 4c^2 \sum_{r,s,t=-\infty}^{\infty} |\kappa(r, s, t)| = \text{O}(K_T/T),$$

where $\text{O}(K_T/T)$ is an expression which has the same asymptotic behaviour as K_T/T as $T \to \infty$.

We remark that q_h is an unbiased estimate for q_g, if the correlation function has period T; this follows by taking the expectation of expression (1). Furthermore the above results imply two corollaries about the consistence of the estimate, i.e. $\lim_{T\to\infty} \text{E}(q_g - q_h) = 0$ and $\lim_{T\to\infty} \text{Var}(q_g - q_h) = 0$. Combining (P1) and (P3) it follows that:

(P5) q_h is a consistent estimate for q_g, if

$$\lim_{T\to\infty} \sum_{s=1}^{T-1} s|g(s)| = 0 \text{ and } \max_{-T<r,s,t<T} |\kappa(r, s, t)|$$

is uniformly bounded for all T.

Combining (P2) and (P4) we have:

(P6) q_h is a consistent estimate for q_g, if the following assumptions are satisfied:

(i) there exists a number $b > 0$, such that $s^b R(s)$ is absolutely summable;

(ii) there exist a number $c > 0$ and a sequence $\{K_T\}$ of integers such that

$K_T \to \infty$, $K_T/T \to 0$ as $T \to \infty$ and $g(s) = 0$ for $K_T < s < T$ and $|g(s)| \leqq c/T$ for $0 \leqq s \leqq K_T$

(iii) $\sum_{r,s,t=-\infty}^{\infty} |\kappa(r, s, t)|$ is finite.

Two examples satisfying (P5) and (P6), respectively, are discussed in Section 4.

3. Computation of a circulant form

3.1. Exactly band limited x

Let

$$X(f)=\sum_{t=0}^{T-1} x(t)w_T^{ft}$$

be the discrete Fourier transform (DFT) of the sequence $x(0), \ldots, x(T-1)$, where $w_T=\exp(-j2\pi/T)$ and H is accordingly the DFT of the sequence $h(0), \ldots, h(T-1)$. A frequency domain representation of the circulant form can be found by substitution:

$$q_h=\sum_{f=0}^{T-1} H(f)|X(f)|^2 \tag{4}$$

which allows the application of a FFT algorithm for a fast computation. Assuming the finite sequence x is exactly narrowband, there exist faster algorithms for calculating the circulant form.[2] For the sake of simplicity we describe only a special version of these algorithms based on quadrature sampling. The others could be used in a similar manner to get corresponding results.

If x is narrowband, we can write

$$x(t)=x_1(t)\cos 2\pi f_0 t/T-x_2(t)\sin 2\pi f_0 t/T$$

where x_1 and x_2 are the quadrature components of x with respect to the central frequency f_0/T and phase 0. The idea was to deduce an interpolation formula for x which uses only the samples $x(an)$ and $x(an+k)$ of x for which $x(an)=x_1(an)$ and $x(an+k)=-x_2(an+k)$ for $n=0, \ldots, \alpha-1$ and calculate the circulant form with support of these samples. The problem was to find suitable integers a, α, and k. In the following we list sufficient conditions. Let

(i) T be even and $x=(x(0), \ldots, x(T-1))$ be band limited, i.e. $X(f)=0$ for f in $\{0, \ldots, T-1\}$ but not in $B=\{f_0-K, \ldots, f_0+K, T-f_0-K, \ldots, T-f_0+K\}$, which is the discrete band with discrete bandwidth $\alpha=2K+1$ and centre f_0;

(ii) 0, $T/2$ be not in B and let there exist integers a and l such that $a\alpha=T$ and $l\alpha=f_0$;

(iii) the integer $0<k<a$ be chosen such that a is neither a submultiple of $2lk$ nor of $(2l+1)k$ and k solves $w_a^{2lk}=-1$.

The algorithm computes the number

$$\hat{q}_h = \frac{1}{\alpha} \sum_{f=0}^{\alpha-1} ((|Z(f)|^2 + |Z(\alpha - f)|^2 H_1(f, 0)/2 \qquad (6)$$
$$+ [\mathrm{Im} Z(f) Z(\alpha - f) - \mathrm{j}(|Z(f)|^2 - |Z(\alpha - f)|^2)/2] H_2(f, k))$$

where Im denotes the imaginary part, and for $f = 0, \ldots, \alpha - 1$

$$Z(f) = \sum_{n=0}^{\alpha-1} (x(an) + \mathrm{j}x(an + k)) w_\alpha^{nf}$$

$$H_1(f, 0) = \frac{a}{4} (\hat{H}_B(f_0 - (\alpha - f)) + \hat{H}_B(f_0 + (\alpha - f)) + \hat{H}_B(f_0 + f) + \hat{H}_B(f_0 - f))$$

$$H_2(f, k) = \mathrm{j} \frac{a}{4} w_T^{fk} ((\hat{H}_B(f_0 - (\alpha - f)) - \hat{H}_B(f_0 + (\alpha - f))] w_a^{-k}$$
$$+ \hat{H}_B(f_0 + f) - \hat{H}_B(f_0 - f))$$

with $\hat{H}_B$ the bandfiltered part of H.

If conditions (5) are satisfied the algorithm computes the exact value of the circulant form, i.e. $\hat{q}_h = q_h$. However, the exact band limitation of the finite sequence x, as assumed in condition (5) (i), is only satisfied in special cases; for example, if $\{x(t)\}$ is a narrowband process with period T. Frequently in applications and especially for large T, the finite observation sequence on a band limited process is approximately band limited. Therefore, we are interested in the error induced by calculating $\hat{q}_h$ instead of q_h in such situations.

A different approximate calculation of q_h under similar conditions is described by Steimel.[6]

3.2. Approximately bandlimited x

Let the process $\{x(t)\}$ have a spectral density $S(v)$ which is zero outside a band contained in

$$\hat{B} = \{2\pi(f_0 - K)/T \leqq v \leqq 2\pi(f_0 + K)/T\} \cup \{2\pi(T - f_0 - K)/T \leqq v$$
$$\leqq 2\pi(T - f_0 + K)/T\}$$

and continuous within the band. $\hat{B}$ corresponds to the discrete band B of subsection A so that the bandwidth of $\{x(t)\}$ is at most $2\pi\alpha/T$. In most cases, the finite sequence x of observations is approximately limited to the band B. Therefore, $\hat{q}_h$ may differ from q_h, since condition (5) (i) is violated. We can prove the following proposition about the bias and variance occurring.

(P7) $\mathrm{E}(\hat{q}_h - q_h) = \mathrm{O}(\max_{0 \leqq f < T} |H(f)| \log a \log T)$ and

$$\mathrm{Var}(\hat{q}_h - q_h) = \mathrm{O}[\max_{0 \leqq f < T} H(f)^2 [(\log a \log T)^2 + \log a (\log \alpha)^3]$$
$$+ T (\log T)^4 \max h(t)^2 \sum_{r,s,t=-\infty}^{\infty} (|\kappa(r, s, t)|)]$$

if conditions (5) are satisfied with the modification that $\{x(t)\}$ is limited to the band $\hat{B}$.

A straightforward, but laborious proof of (P7) is indicated in the Appendix.

Using $\hat{q}_h$ as an estimate for q_h, (P7) implies that the bias and the variance are the smaller, the smaller a is. The most complex calculations needed for the computation of $\hat{q}_h$ by formula (6) are α-point Fourier transforms, if the numbers $H_1(f, 0)$ and $H_2(f, k)$ are calculated *a priori*. Therefore, we need to balance the sampling rate a, which must be large for a fast computation, against the estimate's statistical stability. Finally, we remark that (P7) contains the following special case:

(P8) $\hat{q}_h$ is a consistent estimate for q_h, if the assumptions of (P7) are satisfied, a is fixed, $\lim_{T\to\infty} \max_{0\leqq f<T}|H(f)|(\log T)^{3/2}=0$, and if for non-Gaussian processes $h(t)=O(1/T)$ and

$$\sum_{r,s,t=-\infty}^{\infty} |\kappa(r, s, t)| \text{ is finite.}$$

The two examples discussed in the next section satisfy (P8).

4. Applications

4.1. Estimating the correlation function

A well known estimate for the correlation function $R(s)$ at a single point $s=\tau$ is

$$c(\tau)=c(-\tau)=\frac{1}{T}\sum_{t=0}^{T-\tau-1} x(t)x(t+\tau)$$

When M is the maximum lag number, τ may be one of the numbers $0, \ldots, M-1$. $c(\tau)$ is a consistent estimate for $R(\tau)$, if

$$\lim_{T\to\infty}\frac{1}{T}\sum_{r=-(T-1)}^{T-1} \kappa(\tau, -r, \tau-r)=0$$

and if one of the following conditions holds (see ref. 3).

$$S(v) \text{ is continuous in } [0, 2\pi) \tag{7}$$

$$R(s) \text{ is square summable} \tag{8}$$

The estimate $c(\tau)$ can be interpreted as a Toeplitz form q_g^{τ} with $g(s)=1/(2T)$ for $s=\pm\tau\neq 0$ or $g(s)=1/T$ for $s=\tau=0$ and $g(s)=0$ otherwise. Then we can proceed as described in Sections 2 and 3, i.e. substitute h for g, approximate q_g^{τ} by q_h^{τ} and again q_h^{τ} by $\hat{q}_h^{\tau}$. The estimate has the following property:

(P9) $\hat{q}_h^\tau$ is a consistent estimate for $R(s)$ at the point $s=\tau$, if conditions (5) are satisfied with the modification that $\{x(t)\}$ is limited to the band $\hat{B}$ and $\Sigma_{r,s,t=-\infty}^{\infty}|\kappa(r, s, t)|$ is finite and (7) or (8) holds.

(P9) is a consequence of (P5) and (P8), since $H(f)=O(1/T)$ and $\Sigma_{s=1}^{T-1} s|g(s)| \leqq \tau/2T$.

For estimating the M values $R(0), \ldots, R(M-1)$ we have to calculate the M numbers $\hat{q}_h^0, \ldots, \hat{q}_h^{M-1}$, which can be done simultaneously as indicated in Section 4.3. We note that M should not be greater than the bandwidth α, since the algorithms for calculating $\hat{q}_h$ use only 2α samples of the sequence $x(0), \ldots, x(T-1)$ and values of M greater than α would produce redundant information.

4.2. Estimating the spectral density

Many estimates of the spectral density $S(v)$, $0 \leqq v \leqq 2\pi$, at a single point $v=\lambda$ have the form

$$\hat{S}(\lambda)=\frac{1}{2\pi}\sum_{s=-(T-1)}^{T-1} w(s) \cos \lambda\, s\, c(s)$$

where $w(s)=w(-s)$, $s=0, \ldots, T-1$, are suitable window coefficients. $\hat{S}(\lambda)$ is a consistent estimate for $S(\lambda)$, if the following conditions hold:[3,4]

The coefficients $w(s)$ are of the special form $w(s)=k(s/K_T)$ for $s=0, \pm 1, \ldots, \pm K_T$ and $w(s)=0$ for $s=\pm(K_T+1), \ldots, \pm(T-1)$, where $k(y)$ is an even function, bounded in the interval $[-1, 1]$, continuous as $y=0$, and with $k(0)=1$ and $\{K_T\}$ is a sequence of integers depending on T. There exist positive numbers b and c, such that $\lim_{y\to 0}(1-k(y))/|y|^b=c$, $s^b R(s)$ is absolutely summable and $K_T^b/T\to 0$ for $b\geqq 1$ or $K_T/T\to 0$ for $b<1$ as $T\to\infty$. Finally $\Sigma_{r,s,t=-\infty}^{\infty}|\kappa(r, s, t)|$ is finite. (9)

$\hat{S}(\lambda)$ can be calculated by a Toeplitz form, namely $\hat{S}(\lambda)=q_g^\lambda$ with $q(s)=1/(2\pi T)w(s) \cos \lambda s$ for $s=0, \ldots, T-1$. Calculating the corresponding number $\hat{q}_h^\lambda$, we have:

(P10) $\hat{q}_h^\lambda$ is a consistent estimate for $S(v)$ at the point $v=\lambda$, if conditions (5) and (9) are satisfied with the modification that $\{x(t)\}$ is limited to the band $\hat{B}$ and if in addition $\lim_{T\to\infty} K_T/T(\log T)^{3/2}=0$.

(P10) follows from (P6) and (P8); conditions (9) imply the assumptions of (P6), and (P8) is satisfied, since for $f=0, \ldots, T-1$

$$\sum_{s=0}^{T-1} h(s) \cos 2\pi\, fs/T=\frac{1}{\pi T}\sum_{s=0}^{T-1}\left(1-\frac{s}{T}\right)w\left(\frac{s}{K_T}\right)\cos \lambda s \cos 2\pi fs/T=O(K_T/T)$$

The various estimates for $S(v)$ at a single point $v=\lambda$ depend on the choice of the window coefficients $w(s)$. Whereas the estimates $\hat{S}(\lambda)$ can be compared on the basis of $\int_{-1}^{1}k(y)^2\mathrm{d}y$,[3] we have in addition to consider $\max_{0\leq f<T}|H(f)|$ for comparing our estimators $\hat{q}_h^{\lambda}$, which follows from (P7).

For estimating $S(v)$ at L points $\lambda_1, \ldots, \lambda_L$ the L numbers $\hat{q}_h^{\lambda_1}, \ldots, \hat{q}_h^{\lambda_L}$ have to be calculated simultaneously, as indicated in Section 4.3. Since the process $\{x(t)\}$ is assumed to be limited to the band $\hat{B}$, the frequencies $\lambda_1, \ldots, \lambda_L$ should be contained in $\hat{B}$. As with the number M of Section 4.1 the number L should not be greater than the bandwidth α.

4.3. Computation time

As indicated above, the estimation of the correlation function and the spectral density of our method requires the calculation of several numbers $\{\hat{q}_h^i\}$ $(i=1, \ldots, N)$ all depending on the same observations. In the following we are interested in the time needed for simultaneously computing $\{\hat{q}_h^i\}$ $(i=1, \ldots, N)$. Finally for comparison, we mention the computation time which results from estimating the correlation function by a method frequently used.

Let k^{r}, k^{c}, $k^{\mathrm{r}}_{\mathrm{FFT}}$, and $k^{\mathrm{c}}_{\mathrm{FFT}}$ be proportionality constants associated with the computation time of a real multiplication and a real addition, a complex multiplication and a complex addition, a real FFT, and a complex FFT, respectively, depending on the program and type of computer. For example, the computation time of a real DFT of length T is approximately $k^{\mathrm{r}}_{\mathrm{FFT}}T\log_2 T$. For calculating a number $\hat{q}_h$ we can use the symmetry in formula (6), which yields

$$\hat{q}_h=\frac{1}{\alpha}\sum_{f=1}^{K}\Big([|Z(f)|^2+|Z(\alpha-f)|^2]H_1(f,0)+2\mathrm{Im}Z(f)Z(\alpha-f)\mathrm{Re}H_2(f,k)$$
$$+[|Z(f)|^2-|Z(\alpha-f)|^2]\mathrm{Im}H_2(f,k)\Big)+\frac{1}{\alpha}|Z(0)|^2H_1(0,0) \tag{10}$$

where Re denotes the real part. If the numbers $H_1(f, 0)$ and $H_2(f, k)$ are computed *a priori*, there remains the computation of $Z(f)$, $|Z(f)|^2$, $\mathrm{Im}Z(f)Z(\alpha-f)$ and the summing up with the weights H_1, $\mathrm{Re}H_2$, $\mathrm{Im}H_2$. Altogether we have an amount of computing time $t(\hat{q}_h)$ approximately equal to $k^{\mathrm{c}}_{\mathrm{FFT}}\alpha\log_2\alpha+4{\cdot}5\alpha k^{\mathrm{r}}$. To estimate the correlation function or spectral density at N single points by our method we have to compute N numbers $\{\hat{q}_h^i\}(i=1, \ldots, N)$, i.e. N-times a formula like (10). The numbers $H_1(f, 0)$ and $H_2(f, k)$ vary for the different estimation points, but all can be computed *a priori*. The DFT Z must be computed only once for all N values. This results in a computation time of $t[\{\hat{q}_h^i\}(i=1, \ldots, N)]\simeq k^{\mathrm{c}}_{\mathrm{FFT}}\alpha\log_2\alpha+\frac{3}{2}\alpha Nk^{\mathrm{r}}$.

A frequently used method for estimating $R(s)$ at N points is to section the T-point sequence x into pieces of length $2N$ so that DFTs of length $2N$ have to be computed, (compare ref. 7). This method yields a computation time of $t(R) \cong k^{r}_{\mathrm{FFT}}(T/N+1)2N \log_2 2N + 2Nk^{c}$. If there is a band limitation and the band width α is sufficiently small, our estimate for $R(s)$ is computable faster than the latter as the following example may illustrate. For $T=2^{10}$ and $N=2^5$ one finds $t(R) \cong 12672k^{r}_{\mathrm{FFT}} + 64k^{c}$, whereas we require only $t[\{\hat{q}^{i}_{h}\}(i=1, \ldots, N)] \cong 896k^{c}_{\mathrm{FFT}} + 6144k^{r}$ for a bandwidth of $\alpha = T/8 = 2^7$.

4.4. Concluding remarks

In this paper we described the quadrature sampling version of the Böhme algorithms which requires that $w_a^{2lk} = -1$, such that it is only applicable if a is a multiple of 4 and for $a=4, 12, 20, \ldots$ l and k are odd and for $a=8, 16, 24, \ldots$ the product kl is even. The general version of the algorithms, which is always applicable, is described in Böhme[2] and also in Steime,[8] where further investigations on the subject are also presented. Concerning the estimation of the spectral density more results can be found in Steimel.[9]

APPENDIX

Proof of (P7)

Let X_1 and X_2 denote the α-point DFTs of the sequences $x(an)$ and $x(an+k)$, $n=0, \ldots, \alpha-1$, respectively. If the finite sequence x is approximately limited to the discrete band B, X_1 and X_2 are aliased versions of X, i.e.

$$X_1(f) = \frac{1}{a}\sum_{r=0}^{\alpha-1} X(r\alpha+f) \quad \text{and} \quad X_2(f) = \frac{1}{a}\sum_{r=0}^{\alpha-1} X(r\alpha+f)w_T^{-(r\alpha+f)k}$$

$$(f=0, \ldots, \alpha-1) \tag{11}$$

Substituting equation (11) and equation (6) yeilds

$$\hat{q}_n = \frac{1}{aT}\sum_{f=0}^{\alpha-1}\sum_{r,s=0}^{\alpha-1} X(r\alpha+f)X(s\alpha+f)^{*}C(f, r, s) \tag{12}$$

where * indicates the complex conjugate and

$$C(f, r, s) = (1+w_a^{(s-r)k})H_1(f, 0) + 2w_T^{-(r\alpha+f)k}H_2(f, k)$$

Defining $F^{B}(f) = 1 - F^{R}(f) = 1$ for f in B and equal to 0 otherwise and taking into account that $\hat{q}_h = q_h$, if (5) is satisfied, the expectation of the difference of expressions (12) and (4) is bounded by

$$|\mathrm{E}(\hat{q}_h - q_h)| \leqq \max_{f\in B}|H(f)| \sum_{f=0}^{\alpha-1} \sum_{r,s=0}^{a-1} \left(\frac{1}{T}|\mathrm{E}X(r\alpha+f)X(s\alpha+f)^*| \right. \tag{13}$$

$$\left. \times (2F^B(r\alpha+f)F^R(s\alpha+f)+F^R(r\alpha+f)F^R(s\alpha+f))\right) + \max_{f\notin B}|H(f)| \sum_{f\notin B} \frac{1}{T}\mathrm{E}|X(f)|^2$$

For further calculations we need the formula

$$\frac{1}{T}|\mathrm{E}X(f_1)X(f_2)^*| = \left| \int_{\hat{B}} S(v) \frac{\sin (v-2\pi f_1/T)T/2 \, \sin (v-2\pi f_2/T)T/2}{T \sin (v-2\pi f_1/T)/2 \, \sin (v-2\pi f_2/T)/2} \mathrm{d}v \right|$$

and the known inequality $\sin x \geqq 2/\pi x$ for $0 \leqq x \leqq \pi/2$. Since $\alpha = 2K+1$ and T was assumed to be even, $a = T/\alpha$ must be even. Recalling $l = f_0/\alpha$, we get for the first sum of the right side of equation (13):

$$\sum_{f=0}^{\alpha-1} \sum_{r,s=0}^{a-1} \frac{1}{T}|\mathrm{E}X(r\alpha+f)X(s\alpha+f)^*| 2F^B(r\alpha+f)F^R(s\alpha+f)$$

$$\leqq \frac{8\pi}{T} \max S(v) \left(2 \sum_{f=-K}^{K} \sum_{\substack{r=-a/2+1 \\ \neq \pm l}}^{a/2} \left| \frac{1}{\sin \pi(f_0 - r\alpha)/T} \right| \right.$$

$$\left. + \sum_{\substack{f,f'=-K \\ f\neq f'}}^{K} \sum_{\substack{r=-a/2+1 \\ \neq \pm l}}^{a/2} \left| \frac{1}{T \sin \pi(f_0 - r\alpha + f' - f)/T \sin \pi(f'-f)/T} \right| \right)$$

$$\leqq \frac{8\pi}{T} \max S(v) \left(4\alpha \sum_{\substack{r=1 \\ \neq 2l}}^{a/2} \frac{T}{2\alpha r} + \sum_{\substack{n=-2K \\ \neq 0}}^{2K} \sum_{\substack{r=1 \\ \neq 2l \\ r\alpha+n \leqq T/2}}^{a/2} \frac{T(\alpha - |n|)}{2|n|(r\alpha+n)} \right)$$

$$\leqq 4\pi \max S(v) \left(4(0{\cdot}6 + \log \tfrac{1}{2}a) + 2(0{\cdot}6 + \log \alpha)(0.6 + \log \tfrac{1}{2}a)\right)$$

where we have used $\Sigma_{n=1}^{N} 1/n \cong 0{\cdot}6 + \log N$. Since

$$\int_{\mu}^{\lambda} \frac{\mathrm{d}v}{\sin^2 v} \leqq \left| \frac{1}{\sin \lambda} \right| + \left| \frac{1}{\sin \mu} \right|$$

we get for the third sum in equation (13)

$$\sum_{f\notin B} \frac{1}{T}\mathrm{E}|X(f)|^2$$

$$\leqq \frac{4}{T} \max S(v) \left(\sum_{n=1}^{T/2-f_0-K} + \sum_{n=1}^{T/2+f_0-K+1} \right) \int_{\pi n/T}^{\pi(n+2K)/T} \frac{\mathrm{d}v}{\sin^2 v}$$

$$\leqq 4 \max S(v)(0{\cdot}6 + \log \tfrac{1}{2}T + \log \tfrac{1}{2}a)$$

The second sum in equation (13) can be handled as the first sum, if $r \neq s$ and as the third sum, if $r = s$. Summarizing we find

$$|E(\hat{q}_h - q_h)| \leqq 2 \max_{0 \leqq f < T} |H(f)| \max S(v) \left((20 \cdot 8\pi + 4) \log \tfrac{1}{2} a + (2 \cdot 4\pi + 4) \log \tfrac{1}{2} T + 4\pi \log \tfrac{1}{2} a \left(\log \tfrac{1}{2} a + \log \alpha\right)\right),$$

which is an expression of order $\max|H(f)| \log a \log T$.

The result for the variance of $\hat{q}_h - q_h$ is obtained by bounding the Gaussian part by the same methods as described above and directly showing that the non-Gaussian part is an expression of order $\Sigma_{r,s,t=-\infty}^{\infty} |\kappa(r, s, t)|$ times $T(\log T)^4 \max h(t)^2$.

References

1. J. F. Böhme, "Detektion mit Rechnern". Habilitationsschrift, Informatik Bericht No. 15, Institut für Informatik der Universität Bonn (1975). (In German).
2. J. F. Böhme, "Fast Adaptive Energy Detection of Narrow Band Signals". Conf. on Digital Signal Pressing, Florence, 1978. Preprint.
3. T. W. Anderson, "The Statistical Analysis of Time Series". New York: Wiley, 1971.
4. E. J. Hannan, "Multiple Time Series". New York: Wiley, 1970.
5. J. Pearl, "On coding and filtering stationary signals by discrete Fourier-transform". *IEEE Trans., on Information Theory*, **IT-19**, pp. 229–232 (1973).
6. U. Steimel, "Approximate Berechnung einer Toeplitzform mit Hilfe von Quadratur Sampling". In Informatik Bericht No. 13, Institut für Informatik der Universität Bonn, pp. 19–34. (In German).
7. A. V. Oppenheim, and Schafer, R. W. "Digital Signal Processing". Englewood Cliffs, N. J.: Prentice-Hall, 1975.
8. U. Steimel, "Fast Computation of Toeplitz Forms under Narrowband Conditions with Applications to Statistical Signal Processing". To be published.
9. U. Steimel, "Fast Estimation of Narrowband Spectra". To be published.

A Technique for the Evaluation of Circularly Symmetric Two-dimensional Fourier Transforms and its Application to the Measurement of Ocean Bottom Reflection Coefficients

ALAN V. OPPENHEIM

Research Laboratory of Electronics, Department of Electrical Engineering and Computer Science, Massachusetts Institute of Technology, Cambridge, U.S.A.

GEORGE V. FRISK

Woods Hole Oceanographic Institution, Woods Hole, U.S.A.

DAVID R. MARTINEZ

MIT-WHOI Joint program in Oceanography, Oceanographic Engineering, Woods Hole, U.S.A.

1. Introduction

In a variety of applications, the need arises for the evaluation of the two-dimensional Fourier transform of circularly symmetric functions. For example, as we discuss in more detail is Section 3, for a horizontally stratified ocean bottom illuminated by an acoustic point source, the reflected pressure field and the plane-wave reflection coefficient are circularly symmetric and related through a two-dimensional Fourier transform. Applying the Fourier transform to the measure field, the plane-wave reflection coefficient can thus be calculated. Other examples arise in such areas as optics and molecular biology.

Since the Fourier transform of a circularly symmetric function is also circularly symmetric, both are specified by a radial slice. The method proposed in this paper for computing the transform is to exploit the "projection-slice" theorem for two-dimensional transforms. In essence, this theorem states that a slice through a two-dimensional transform is the one-dimensional transform of a projection of the original two-dimensional function. In the case of a circularly symmetric function, the one-dimensional transform of a single projection specifies the entire two-dimensional transform. In Section 2 of this paper this basic approach is developed. In Section 3 we consider its specific application to the measurement of the plane wave reflection coefficient of the ocean bottom.

2. Evaluation of the Fourier transform of circularly symmetric functions[1]

Let $f(x, y)$ and $F(\mu, \nu)$ denote a two-dimensional function and its Fourier transform so that

$$F(\mu, \nu)=\frac{1}{2\pi}\int_{-\infty}^{+\infty}\int_{-\infty}^{+\infty} f(x, y)\mathrm{e}^{\mathrm{i}\mu x}\,\mathrm{e}^{\mathrm{i}\nu y}\,\mathrm{d}x\mathrm{d}y$$

or, with $f(x, y)$ and $F(\mu, \nu)$ expressed in polar coordinates,

$$\mathcal{F}(\rho, \phi)=\frac{1}{2\pi}\int_{0}^{2\pi}\int_{0}^{\infty} \not{f}(r, \theta)\mathrm{e}^{\mathrm{i}[\cos(\theta-\phi)]r\rho} r\mathrm{d}r\mathrm{d}\theta \tag{1}$$

where θ is measured relative to the x-axis and ϕ is measured relative to the μ-axis. If $\not{f}(r, \theta)$ is circularly symmetric and thus of the form

$$\not{f}(r, \theta)=g(r)$$

then $\mathcal{F}(\rho, \theta)$ is likewise circularly symmetric

$$\mathcal{F}(\rho, \theta)=G(\rho)$$

and equation (1) reduces to

$$G(\rho)=\int_{0}^{\infty} J_0(r\rho)g(r)r\mathrm{d}r \tag{2}$$

where $J_0(\cdot)$ is the zero-order Bessel function. $g(r)$ and $G(\rho)$ are a radial slice through $f(x, y)$ and $F(\mu, \nu)$ respectively. The integral relationship of equation (2) is commonly referred to as the Hankel or Fourier–Bessel transform.

Our proposed method of numerically evaluating equation (2) is based on the "projection-slice" theorem for the two-dimensional Fourier transform. This theorem states that the one-dimensional transform of a *projection* of $f(x, y)$ at any angle is a *slice* at the same angle of $F(\mu, \nu)$. For example, let us consider the slice in $F(\mu, \nu)$ corresponding to $\nu=0$ or equivalently $\mathcal{F}(\rho, \phi)$ for $\phi=0$. Then

$$F(\mu, 0)=\frac{1}{2\pi}\int_{-\infty}^{+\infty} \mathrm{e}^{\mathrm{i}\mu x}\, p(x)\mathrm{d}x \tag{3}$$

where

$$p(x)=\int_{-\infty}^{+\infty} f(x, y)\mathrm{d}y \tag{4}$$

is the projection of $f(x, y)$ onto the x-axis. Thus, from equation (3), we can write that

$$G(\rho)=\frac{1}{2\pi}\int_{-\infty}^{+\infty} e^{i\rho x}\, p(x)dx \tag{5}$$

Thus, it follows that the Hankel transform can be equivalently expressed (and calculated) as the one-dimensional Fourier transform of the projection $p(x)$.

The two basic computational steps in evaluating equation (2) using this approach are the evaluation of the projection $p(x)$ (equation (4)) and the evaluation of the one-dimensional Fourier transform (equation (5)). Let us assume that $G(\rho)=0$, $|\rho|\geqq R$. Then, from equation (4), $p(x)$ is band limited, and consequently, by virtue of the sampling theorem,

$$G(\rho)=\frac{\Delta x}{2\pi}\sum_{k=-\infty}^{+\infty} p(k\Delta x)\ e^{i\rho k\Delta x} \tag{6}$$

provided that $\Delta x<\pi/R_0$. If we consider calculating $G(\rho)$ at N equally spaced values $\Delta\rho=(1/N)(2\pi/\Delta x)$, then

$$G(k\Delta\rho)=\frac{\Delta x}{2\pi}\sum_{n=0}^{N-1}\left[\sum_{r=-\infty}^{+\infty} p[n+rN)\Delta x]\right] e^{i(2\pi/N)nk} \tag{7}$$

Thus, $G(k\Delta\rho)$, $k=0, 1, \ldots, N-1$ is proportional to the discrete Fourier transform of the samples of $p(x)$, aliased in x. If the samples of $p(x)$ represent a finite length sequence of length $\leqq(N\Delta x)$, then equation (7) reduces to

$$G(k\Delta\rho)=\frac{\Delta x}{2\pi}\sum_{n=0}^{N-1} p(n\Delta x)\ e^{i(2\pi/N)nk} \tag{8}$$

Both equations (7) and (8) correspond to the discrete Fourier transformed and consequently they can be evaluated directly using the one-dimensional FFT.

The calculation of samples of $p(x)$ is somewhat less direct. Equation (4) can equivalently be written as

$$p(x)=2\int_{0}^{+\infty} g(x^2+y^2)^{1/2}dy \tag{9a}$$

$$p(x)=2\int_{x}^{+\infty} g(r)\frac{r}{(r^2-x^2)^{1/2}}dr \tag{9b}$$

$$p(x)=2x\int_{0}^{\pi/2} g\left(\frac{x}{\cos\theta}\right)d\theta \tag{9c}$$

Equations (9) incorporate the fact that since f (r, θ) is conjugate antisymmetric in θ, only its real part contributes to $p(x)$. We have found it most convenient to calculate $p(x)$ through the use of equation (9*a*). Specifically, we note that since $f(x, y)$ is band limited,

$$\int_{-\infty}^{+\infty} f(x, y)\,dy = \Delta y \sum_{k=-\infty}^{+\infty} f(x, k\Delta y) \tag{10}$$

provided only that $\Delta y < 2\pi/R_0$. Equation (10) is basically a consequence of the fact that for a band limited function sampled at one-half the Nyquist rate or higher, its integral is directly proportional to the sum of its samples. Thus, $p(n\Delta x)$ as required in equations (7) or (8) is

$$p(n\Delta x) = \Delta y \sum_{k=-\infty}^{+\infty} g[(n\Delta x)^2 + (k\Delta y)^2]^{1/2} \tag{11}$$

Equations (7) and (11) together provide an exact expression for the numerical calculation of $G(k\Delta\rho)$ provided only that $G(\rho) = 0$, $|\rho| > R_0$. If this is not the case, then equation (7) will compute samples of $G(\rho)$ aliased in ρ, i.e.

$$\sum_{q=-\infty}^{+\infty} G[\Delta\rho(k + qN)] \tag{12}$$

and an integration rule more complex than equation (10) must be used to calculate $p(x)$.

To evaluate equation (11), we assume that $g(x^2 + y^2)^{1/2}$ is known on a rectangular grid in the x–y plane. In many practical cases of interest, $g(r)$ is generally available as samples in r. In this case evaluation of equation (11) requires an interpolation of samples of $g(r)$ to the sample points on the rectangular grid. Under the assumption that $G(\rho) = 0$, $|\rho| > R_0$ this is the only step in the procedure in which an approximation is required.

3. Measurement of the plane-wave reflection coefficient of the ocean bottom[2]

The problem of estimating the plane-wave reflection coefficient of a horizontally stratified ocean bottom has received considerable attention. In general, the reflection coefficient is a function of frequency and incident angle and knowledge of this function provides the information required for the solution of acoustic problems in the water column. One common technique for measurement of the reflection coefficient is to utilize the geometrical acoustics approximation that for a point source and receiver sufficiently far from the bottom, the bottom reflected signal is a spherical wave emanating from the image source and multiplied by the reflection coefficient evaluated at the specular angle.[3] As the specular angle is varied by changing the source–receiver geometry, the reflection coefficient as a

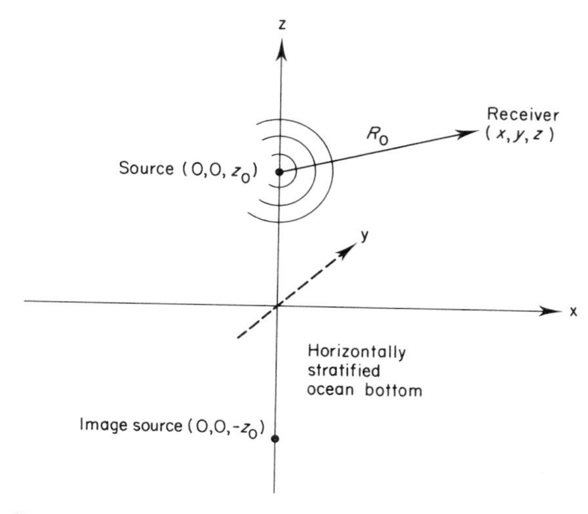

Fig. 1. Geometry for reflection from a horizontally stratified ocean bottom.

function of angle can then be measured. The primary limitation of this method is the assumption of specular reflection, which even for simple bottom types is not valid in the entire specular angle domain.[4] In addition, the conventional method is limited to real angles of incidence between 0 and $\frac{1}{2}\pi$. In comparison, the method which we outline below provides information about the reflection coefficient for both real and complex angles of incidence, corresponding to the reflection of both pure and inhomogeneous plane waves.[6]

The geometry is indicated in Fig. 1 where a point acoustic source at a fixed frequency ω is located at $(0, 0, z_0)$. The incident pressure field p_i at an observation point (x, y, z) is given (with an $\exp(-\mathrm{i}wt)$ time-dependence suppressed) by

$$p_i(x, y, z) = \frac{\mathrm{e}^{ikR_0}}{R_0} \tag{13}$$

where $R_0 = [x^2 + y^2 + (z - z_0)^2]^{1/2}$ and $k = \omega/c$, with c denoting the sound speed. The spherical wave can be decomposed into plane waves by Fourier transforming in the wavenumbers k_x and k_y:[5]

$$p_i(x, y, z) = \frac{i}{2\pi} \iint_{-\infty}^{\infty} \frac{\mathrm{d}k_x \mathrm{d}k_y}{k_z} \exp\{\mathrm{i}[k_x x + k_y y + k_z|z - z_0|]\} \tag{14}$$

with

$$k^2 = k_x^2 + k_y^2 + k_z^2 \tag{15}$$

While the wavenumber components k_x and k_y in the x and y directions respectively are always real, k_z takes on both real and imaginary values by virtue of equation (15). For real values of k_z the plane waves in equation (14) are pure plane waves propagating in the direction (k_x, k_y, k_z). For imaginary values of k_z the plane waves are *inhomogeneous* waves which propagate in the (k_x, k_y) direction and attenuate exponentially as $\exp[-|k_z(z-z_0)|]$ in the z-direction. The reflected pressure field can also be decomposed into plane waves, each one emanating from the image source at $(0, 0, -z_0)$ and multiplied by the plane wave reflection coefficient:

$$p_r(x, y, z)=\frac{i}{2\pi}\iint_{-\infty}^{\infty}\frac{dk_x dk_y}{k_z}R(k_x, k_y, k_z)\exp\{i[k_x x+k_y y+k_z(z+z_0)]\} \quad (16)$$

Now, let us consider measuring the pressure field and assume that $p_r(x, y, z)$ alone can be identified, for example, by using a time-limited pulse or by subtracting out the incident field. With the pressure field measured in the x–y plane at a specified height z, we express $p_r(x, y, z)$ as $p_r(x, y; z)$ to identify the receiver height z as a parameter rather than a variable. Since we assume k to be fixed and since k_x, k_y and k_z are constrained by equation (15), it is convenient to express the reflection coefficient as $R(k_x, k_y; k)$ which also identifies k as a parameter but not as a variable. With this modified notation, equation (16) becomes

$$p_r(x, y; z)=\frac{i}{2\pi}\iint_{-\infty}^{+\infty}\frac{dk_x dk_y}{k_z}R(k_x, k_y; k)\exp\{i[k_x x+k_y y+k_z(z+z_0)]\} \quad (17)$$

Equation (17) is in the form of a two-dimensional spatial-wave number Fourier transform, i.e.

$$p_r(x, y; z)\overset{\mathcal{F}}{\longleftrightarrow}\frac{i}{k_z}R(k_x, k_y; k)\exp[ik_z(z+z_0)] \quad (18)$$

where $\overset{\mathcal{F}}{\longleftrightarrow}$ denotes a two-dimensional Fourier transform pair. Thus,

$$R(k_x, k_y; k)=\frac{-ik_z}{2\pi}\exp[-k_z(z+z_0)]\iint_{-\infty}^{+\infty}dx dy\, p_r(x, y; z)\exp[-i(k_x x+k_y y)] \quad (19)$$

The double integral in equation (19) is the inverse two-dimensional Fourier transform of the measured pressure field.

For a horizontally stratified ocean bottom, the reflection coefficient is

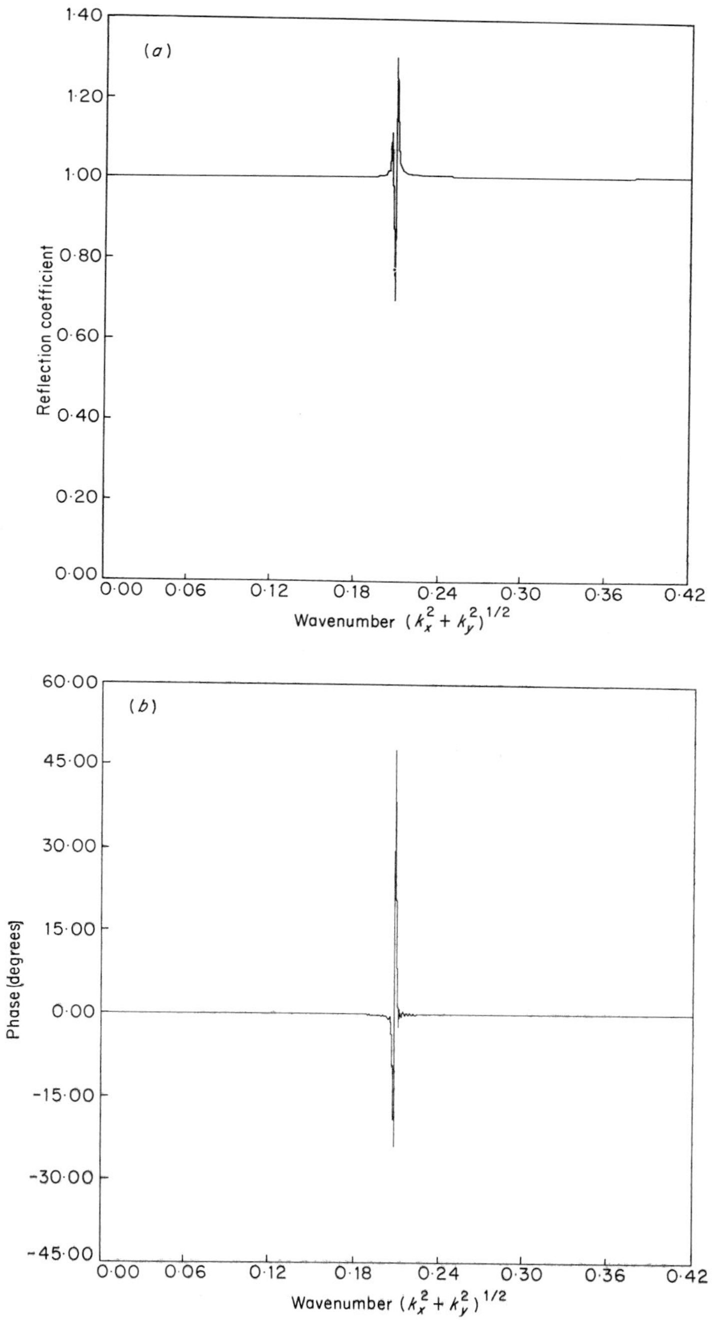

Fig. 2. Computed reflection coefficient for a perfectly reflecting ocean bottom: (*a*) magnitude; (*b*) phase.

circularly symmetric in the (k_x, k_y) plane and consequently the reflected pressure field $p_r(x, y; z)$ is circularly symmetric in the (x, y) plane. Thus, evaluation of the inverse transform in equation (19) can be carried out using the procedure discussed in Section 2.

As an example, we have considered the case of a perfectly reflecting hard bottom in which case the reflection coefficient has unity magnitude and zero phase, and the reflected pressure field is given exactly by

$$p_r(x, y; z) = \frac{\exp[ik(x^2+y^2+(z+z_0)^2)^{1/2}]}{[x^2+y^2+(z+z_0)^2]^{1/2}} \tag{20}$$

Thus, $p_r(x, y; z)$ as given in equation (20) was used in equation (19) with the double integral evaluated using the procedure outlined in Section 2. The values assumed for the parameters were:

$\omega = 100\pi$ rad s^{-1} ($f = 50$Hz);
$c = 1500$ m s^{-1};
$z = z_0 = 10$ m.

The pressure field was assumed band limited with maximum wavenumbers $k_x = 0{\cdot}6$ and $k_y = 0{\cdot}3$ and in computing the projections, the pressure field was sampled accordingly. A cylindrical Hanning window of diameter 5361·65 m was applied to $p_r(x, y; z)$ when the projections were computed.

Since $R(k_x, k_y; k)$ is circularly symmetric, it is a function of $(k_x^2+k_y^2)^{1/2}$. The computed magnitude and phase of the reflection coefficient are shown in Fig. 2 over a range of values of $(k_x^2+k_y^2)^{1/2}$ which includes both pure and inhomogeneous plane waves. While the result compares favourably with the true reflection coefficient $(R=1)$ the effect due to the singularity introduced as k_z becomes imaginary is evident. In a later paper a more detailed treatment of this effect and the effect of windowing the pressure field will be considered.

Acknowledgment

This work was supported in part by the Advanced Research Projects Agency, monitored by ONR under Contract N00014-75-C-0951-NR049-308 and in part by ONR Contract N00014-77-C-0196.

References

1. A. V. Oppenheim, G. V. Frisk and D. R. Martinez, "An Algorithm for the Numerical Evaluation of the Hankel Transform," *Proc. IEEE*, **66**, pp. 264–265 (1978).
2. G. V. Frisk, A. V. Oppenheim, and D. R. Martinez, "Technique for Measuring

the Plane-Wave Reflection Coefficient of the Ocean Bottom," *J. Acoustical Society of America*, **62**, S66 (A) (1977).

3. O. F. Hastrup, "Digital Analysis of Acoustic Reflectivity in the Tyrrhenian Abyssal Plain," *J. Acoustical Society of America*, **47**, pp. 181–190 (1970).
4. D. C. Stickler, "Negative Bottom Loss, Critical-Angle Shift, and the Interpretation of the Bottom Reflection Coefficients," *J. Acoustical Society of America*, **61**, pp. 707–710 (1977).
5. L. M. Brekhovskikh, "Waves in Layered Media." New York: Academic Press (1960).
6. G. V. Frisk, "Inhomogeneous Waves and the Plane Wave Reflection Coefficient," to be published.

A Two-dimensional Fast Fourier Transform for Hexagonally Sampled Data

RUSSELL M. MERSEREAU

School of Electrical Engineering, Georgia Institute of Technology, Atlanta, U.S.A.

1. Introduction

Although there are a number of means by which two-dimensional wave forms can be sampled for digital processing, rectangular sampling is almost universally used. Among the alternative strategies, hexagonal sampling, discussed by Petersen and Middleton,[1] would appear to possess several decided advantages. Mersereau[2] has demonstrated that for waveforms which are bandlimited with a circularly symmetric region of support in the Fourier plane, hexagonal sampling affords computational savings of up to 50% and savings in storage of roughly 25% over the comparable rectangular operations. In this item we shall present an algorithm for performing a 2-D discrete Fourier transform (DFT) on a hexagonal data array. This algorithm, which represents a variation of Rivard's algorithm[3,4] for rectangular arrays, involves 25% less computation and storage than that algorithm for comparably-sized transforms. This corresponds to roughly 42% less computation than that required for a rectangular row–column FFT.

2. Hexagonally sampled signals

Figure 1(*a*) shows the raster of sampling locations that defines a hexagonal array and Fig. 1(*b*) shows the assumed region of support for the Fourier transform of the sampled signal. We shall assume that this region of support is a regular hexagon. The theory has been generalized to consider arbitrary hexagonal regions of support, but that generality is not needed here.

Hexagonal sequences can be determined from their Fourier transforms, which can most conveniently be defined by

$$X(\omega_1, \omega_2) = \sum_{n_1}\sum_{n_2} x(n_1, n_2)\exp\left[-j\left(\frac{2\omega_1}{\sqrt{3}}n_1 + \left(\omega_2 - \frac{\omega_1}{\sqrt{3}}\right)n_2\right]\right. \quad (1)$$

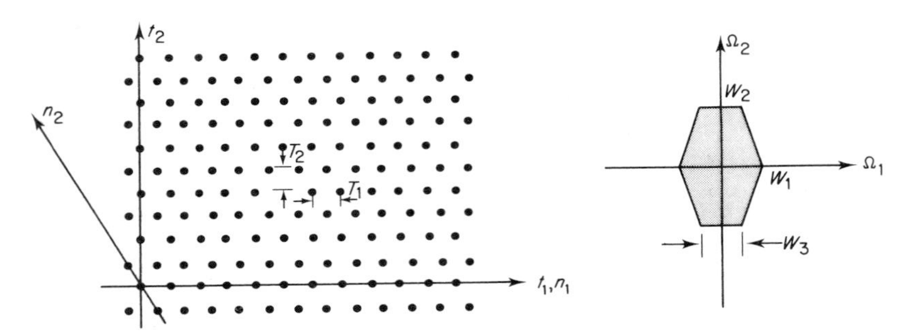

Fig. 1. A hexagonal sampling raster and its hexagonal region of support in the Fourier plane. For exact reconstruction $T_1 < 4\pi/(2W_1 + W_3)$, $T_2 < \pi/W_2$.

$$x(n_1, n_2) = \frac{1}{2\sqrt{3\pi^2}} \iint_{\text{hexagon}} X(\omega_1, \omega_2) \exp\left[j\left(\frac{2n_1 - n_2}{\sqrt{3}} \omega_1 + n_2 \omega_2 \right) \right] d\omega_1 \, d\omega_2 \tag{2}$$

$X(\omega_1, \omega_2)$ is a hexagonally periodic sequence; this periodicity is identical to that observed if a plane is covered with regular hexagonal tiles. Specifically $X(\omega_1, \omega_2)$ is unchanged if the following replacements are made:

$$\begin{gathered} \omega_2 \to \omega_2 + 2\pi \qquad \omega_1 \to \omega_1 + 2\pi\sqrt{3} \\ \omega_1 \to \omega_1 \pi\sqrt{3},\ \omega_2 \to \omega_2 + \pi \qquad \omega_1 \to \omega_1 - \pi\sqrt{3},\ \omega_2 \to \omega_2 + \pi \end{gathered} \tag{3}$$

The integral in equation (2) is performed over one hexagonal period of $X(\omega_1, \omega_2)$.

3. Hexagonally periodic sequences

Consider now a sequence which is of finite extent. In particular let us assume that

$$x(n_1, n_2) = 0, \qquad \begin{aligned} & n_1 < 0,\ n_1 \geqq 2N \\ & n_2 < 0,\ n_2 \geqq 2N \\ & n_1 - n_2 \leqq -N,\ n_1 - n_2 \geqq N \end{aligned} \tag{4}$$

This sequence contains $3N^2$ nonzero samples. It can be modelled as one period of a 2-D hexagonally periodic sequence. If $\tilde{x}(n_1, n_2)$ represents the (hexagonally) periodic extension of $x(n_1, n_2)$ then

$$\begin{aligned} x(n_1, n_2) = x(n_1 + 3N, n_2) &= x(n_1, n_2 + 2N) \\ &= x(n_1 + N, n_2 + N) \\ &= x(n_1 + N, n_2 - N) \end{aligned} \tag{5}$$

In Fig. 2 we show $x(n_1 n_2)$, $\tilde{x}(n_1, n_2)$, and another representation of $\tilde{x}(n_1, n_2)$ for which the periods are shaped like parallelograms.

The periodic sequence $\tilde{x}(n_1, n_2)$ can be represented exactly by a (hexagonal) Fourier series. The coefficients of this series correspond to a hexagonal sampling of the Fourier transform $X(\omega_1, \omega_2)$. This allows us to define the hexagonal DFT for the sequence $x(n_1, n_2)$ as

$$X(k_1, k_2) = \sum_{n_1}\sum_{n_2} x(n_1, n_2) \exp\left[-j\pi\left(\frac{2n_1 - n_2}{3N}(2k_1 - k_2) + \frac{k_2 n_2}{N}\right)\right] \tag{6}$$

$$x(n_1, n_2) = \frac{1}{3N^2}\sum_{k_1}\sum_{k_2} X(k_1, k_2) \exp\left[j\pi\left(\frac{2n_1 - n_2}{3N}(2k_1 - k_2) + \frac{k_2 n_2}{N}\right)\right] \tag{7}$$

$X(k_1, k_2)$ so defined is a hexagonally periodic sequence and the sum in equation (7) is evaluated over one period of that sequence. Actually, performing the sum over a hexagonally-shaped region is awkward computationally. Because of the periodicity of both $\tilde{x}(n_1, n_2)$ and $\tilde{X}(k_1, k_2)$, however, these sums can be evaluated over the regions $0 \leqq n_1 \leqq 3N-1$, $0 \leqq n_2 \leqq N-1$ and $0 \leqq k_1 \leqq 3N-1$, $0 \leqq k_2 \leqq N-1$, respectively.

The hexagonal DFT serves the same role for hexagonal systems that the rectangular DFT does for rectangular systems. Since it represents samples of the Fourier transform it is useful for spectral analysis or for the implementation of FIR filters. This DFT involves $3N^2$ complex sample values, where N can be interpreted as the radius of the finite array. The rectangular DFT with comparable frequency resolution requires $4N^2$ complex sample values. Thus one advantage of the hexagonal DFT over the rectangular one is immediately apparent; it requires 25% less storage. In the next section, it will be shown that it involves less computation as well.

4. Evaluation of the hexagonal DFT

Assume that N is a multiple of 2. Then the sum in equation (6) can be decomposed into four sums—one over the even-indexed samples of both n_1 and n_2, one for those samples for which n_1 is even and n_2 is odd, etc. Then

$$X(k_1, k_2) = S_1(k_1, k_2) + S_2(k_1, k_2) + S_3(k_1, k_2) + S_4(k_1, k_2) \tag{8}$$

where

$$S_1(k_1, k_2) = \mathrm{DFT}_{N/2}[x(2r, 2s)] \tag{9}$$

$$S_2(k_1, k_2) = \exp\left[-j\frac{2\pi}{3N}(k_1 - 2k_2)\right]\mathrm{DFT}_{N/2}[x(2r, 2s+1)] \tag{10}$$

$$S_3(k_1, k_2) = \exp\left[-j\frac{2\pi}{3N}(2k_1 - k_2)\right]\mathrm{DFT}_{N/2}[x(2r+1, 2s)] \tag{11}$$

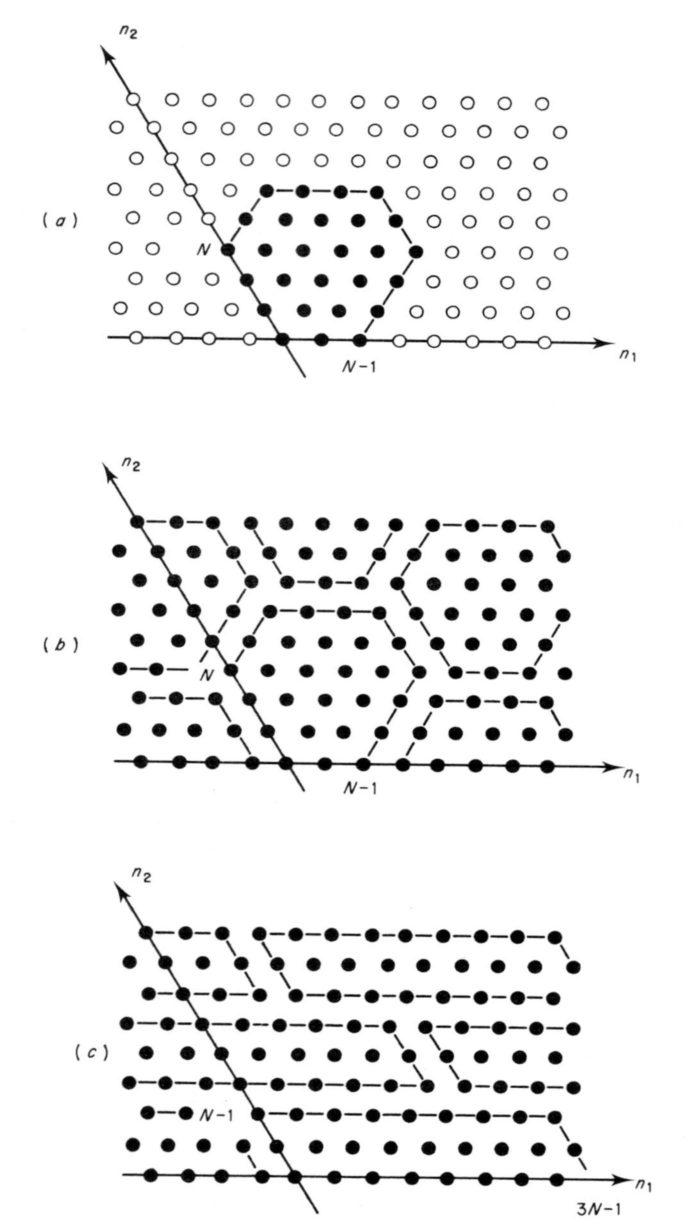

Fig. 2. (*a*) A finite area array. (*b*) The periodic extension of that array demonstrating hexagonal periodicity. (*c*) Another periodic extension of the finite array where the fundamental periods are parallelograms.

$$S_4(k_1, k_2)=\exp\left[-j\frac{2\pi}{3N}(k_1+k_2)\right]\mathrm{DFT}_{N/2}\left[x(2r+1, 2s+1)\right] \qquad (12)$$

The algebraic steps involved in obtaining equations (9–12) have been omitted but are analogous to those required for the one-dimensional case. It should also be noted that $S_i(k_1, k_2)$ $(i=1, 2, 3, 4)$ is hexagonally periodic as described by equation (5) with N replaced by $N/2$. Thus equations (9–12) can be evaluated for all values of k_1, k_2 from the $3(N/2)^2$ values output from the smaller DFTs. In particular, if F, G, H, and I denote the $3(N/2)^2$ point DFTs in equations (9–12) respectively, then

$$X(k_1, k_2)=F(k_1, k_2)+W_{3N}^{2k_2-k_1}G(k_1k_2)+W_{3N}^{2k_1-k_2}H(k_1, k_2) + W_{3N}^{k_1+k_2}I(k_1,k_2)$$

$$X\left(k_1+\frac{3N}{2}, k_2\right)=F(k_1, k_2)-W_{3N}^{2k_2-k_1}G(k_1, k_2)+W_{3N}^{2k_1-k_2}H(k_1, k_2) - W_{3N}^{k_1+k_2}I(k_1, k_2)$$

$$X(k_1+N, k_2+N/2)=F(k_1, k_2)+W_{3N}^{2k_2-k_1}G(k_1, k_2)-W_{3N}^{2k_1-k_2}H(k_1, k_2) - W_{3N}^{k_1+k_2}I(k_1,k_2)$$

$$X\left(k_1+\frac{5N}{2}, k_2+N/2\right)=F(k_1, k_2)-W_{3N}^{2k_2-k_1}G(k_1, k_2)-W_{3N}^{2k_1-k_2}H(k_1, k_2) + W_{3N}^{k_1+k_2}I(k_1, k_2)$$

Thus the computation of the hexagonal coefficients requires $9N^2/16$ complex multiplications in addition to those required to compute the arrays F, G, H and I. If $N=2^\mu$, we can continue this decomposition μ times as with the 1-D DFT. At the end we must perform N 1-D radix-3 butterflies. Each of these can be performed using eight real multiplies. A flowchart for an $N=2$ hexagonal DFT is shown in Fig. 3.

5. Computational savings using a hexagonal DFT

The algorithm as described requires $8N^2\log_2 N+8N^2$ real multiplications and $3N^2$ (complex) storage locations. By way of comparison, Rivard's algorithm applied to a $2N\times 2N$ point rectangular array requires $12N^2\log_2 N+12N^2$ real multiplications and $4N^2$ storage locations. The traditional method of computing a rectangular DFT by computing 1-D DFTs on the rows and columns of the 2-D array requires $16N^2\log_2 N+16N^2$ multiplications and $4N^2$ complex storage locations. Thus the hexagonal

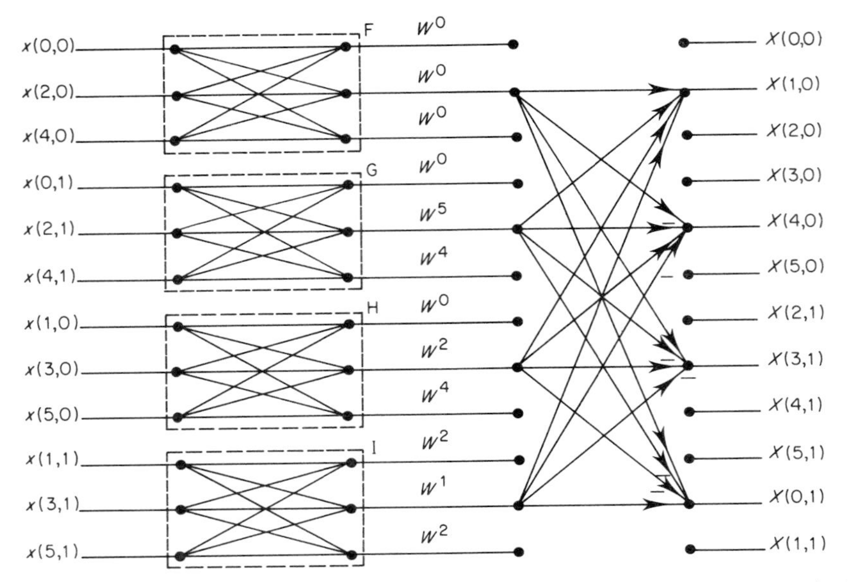

Fig. 3. A flow chart for an $N=2$, twelve-point hexagonal DFT. The coefficients are omitted from the radix-3 butterflies and only one of the four butterflies in the second stage is shown.

DFT saves more than 25% computation and 25% storage over the most efficient of the rectangular algorithms.

Acknowledgment

This research was supported in part by the U.S. Army Research Office under Contract DAAG29-78-C-0005.

References

1. D. P. Petersen and D. Middleton, "Sampling and Reconstruction of Wave-Number Limited Functions in N-Dimensional Euclidean Spaces," *Information and Control*, **5**, pp. 279–323 (1962).
2. R. M. Mersereau, "Two-Dimensional Signal Processing from Hexagonal Rasters," 1978 IEEE Int. Conf. on Acoust., Speech, and Signal Processing, *Record*, pp. 739–742 (1978).
3. G. E. Rivard, "Direct Fast Fourier Transform of Bivariate Functions," *IEEE Trans. Acoustics, Speech and Signal Processing*, **ASSP-25**, pp. 250–252 (1977).
4. D. B. Harris, J. H. McClellan, D. S. K. Chan, and H. W. Schuessler, "Vector Radix Fast Fourier Transform," 1977 IEEE Int. Conf. on ASSP *Record*, pp. 548–551 (1977).

Part 3
IMPLEMENTATIONS

The Application of Finite Arithmetic Structures to the Design of Digital Processing Systems

P. J. W. RAYNER

Department of Engineering, Cambridge University, England

1. Introduction

Most digital signal processing systems are required to operate on variables which are derived from some continuous physical process. In order to achieve this the signal parameters must be quantized in time and magnitude. Quantization in time is, of course, well understood and is achieved by sampling the continuous variable at discrete instants; the effects of sampling in time are contained in the sampling theorem. Quantization in amplitude is usually achieved by representing the amplitude of each sample from the continuous signal by the "closest" number from the set of numbers representable in the digital processing system. The amplitude quantization process introduces an error which may, in many cases, be considered as a random variable which is added to the exact representation of the signal.

The algebraic description of digital signal processing systems is not usually in terms of quantized amplitude variables but is specified by equations in terms of continuous amplitude variables sampled in time. Many analytical techniques (e.g. z-transform) are available for the analysis of discrete time signals. An example of a digital processing system described in terms of discrete-time continuous-amplitude variables is the first-order recursive digital filter shown in Fig. 1(a).

The realization of such a system is to be achieved in terms of digital circuits which operate on only a finite set of values of the variables. This is normally achieved by simply approximating the continuous arithmetic operations in the algebraic description of the system by discrete arithmetic operations. The realization of the first-order filter is shown in Fig. 1(b), where the multiplier and adder are digital circuits, usually binary, which approximate to the continuous algebraic operations. The multiplication of two m-bit binary numbers produces, in general, a $2m$-bit result which must be approximated in some way by an n-bit number; truncation or rounding

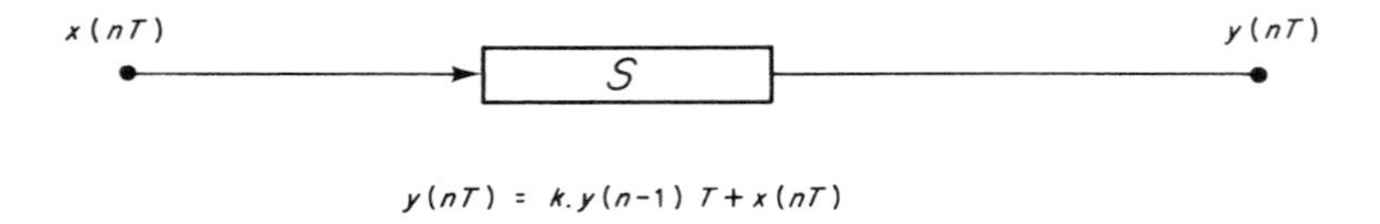

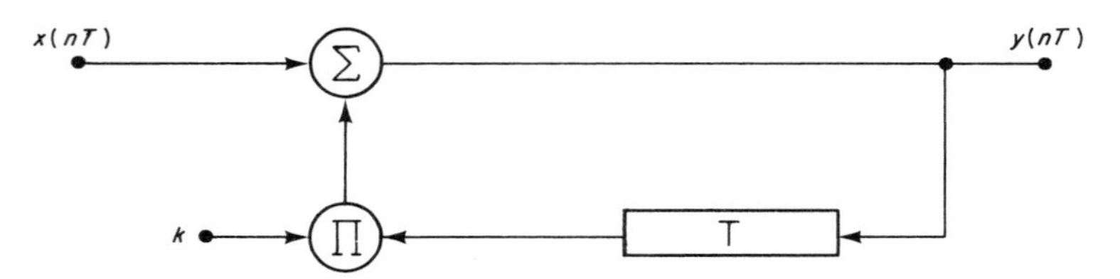

Fig. 1. (*a*) First-order recursive filter. (*b*) Realization of first-order recursive filter.

is the most common technique for achieving this end. In even a relatively simple system it is difficult to predict the effects of these non-linear operations on the overall system behaviour. Usually the best that can be done is a statistical analysis of the errors. In systems with feedback (e.g. recursive filters) the nonlinearity introduced by truncation or rounding can produce limit-cycle oscillations.

One must consider whether these effects are a fundamental limitation of digital processing techniques or whether they are produced as a result of limitations in the synthesis techniques. In general the operation of digital system is that of mapping k input variables, each defined to m-bits, into j output variables, also usually defined to m-bits. However the continuous algebraic system description does not have this form so it might be argued that the system description is at fault. One method of overcoming this problem is to calculate the required mapping directly from the continuous system equations.

2. Boolean equation realization

A digital processing system is a finite-state machine and has, therefore, a finite number of possible values of inputs and states. In principle it is possible to calculate the system output from the continuous system equations for every possible combination of the discrete inputs and states. In general the calculated values of output will not be representable in the system so that they must be approximated by values which are representable. A simple example is shown in Table 1 where an exhaustive listing is made of all possible input values $\bar{x}_n$ and state values $\bar{y}_n$ for a first-order recursive filter in which

$$\bar{x}_n,\ \bar{y}_{n-1} \in I = \{i:\ 0 \leqq i \leqq 3\}$$

TABLE 1. $y_n = 0{\cdot}9\ \bar{y}_{n-1} + \bar{x}_n$

$$\bar{y}_n = \begin{cases} y_n \text{ truncated, } y < 4 \\ 3,\ y_n \geqq 4 \end{cases}$$

$\bar{x}_n$	$\bar{y}_{n-1}$	y_n	$\bar{y}_n$
0	0	0	0
0	1	0·9	0
0	2	1·8	1
0	3	2·7	2
1	0	1·0	1
1	1	1·9	1
1	2	2·8	2
1	3	3·7	3
2	0	2·0	2
2	1	2·9	2
2	2	3·8	3
2	3	4·7	3
3	0	3·0	3
3	1	3·9	3
3	2	4·8	3
3	3	5·7	3

The corresponding output values y_n are calculated from the continuous filter equation:

$$y_n = 0{\cdot}9\bar{y}_{n-1} + \bar{x}_n\ .$$

The output values y_n must be approximated by $\bar{y}_n \in I$; in the example, truncation and saturation has been applied to give the values $\bar{y}_n$. The relationship between the variables $\bar{x}_n$, $\bar{y}_{n-1}$ and $\bar{y}_n$ defined by Table 1 is entirely in terms of discrete amplitude variables and may be realized by digital circuits. Moreover the filter output y_n is always within one bit of the output y_n calculated from the continuous filter equation; apart from saturation. The relationships of Table 1 may be rewritten in terms of binary variables to produce a Boolean truth table (Table 2). Conventional minimization techniques (e.g. Quine–McCluskey) may be applied to produce a minimal logic realization of the system.

The advantages of this technique are that the output of the system is always within one bit of the output defined by the continuous system equations and the logical realization contains the minimum number of gates and is therefore a potentially high-speed realization. This should be contrasted with the direct method of realization where the algebraic operations in the continuous system equations are simply approximated by the corresponding discrete algebraic operations with no guarantee that the output of the system so realized conforms very closely to the required system.

TABLE 2. Binary version of Table 1.
$e = a + c(b + d)$
$F = (\bar{a} + c)(a + c + b \oplus d)$.

$\bar{x}_n$		$\bar{y}_{n-1}$		$\bar{y}_n$	
a	b	c	d	e	f
0	0	0	0	0	0
0	0	0	1	0	0
0	0	1	0	0	1
0	0	1	1	1	0
0	1	0	0	0	1
0	1	0	1	0	1
0	1	1	0	1	0
0	1	1	1	1	1
1	0	0	0	1	0
1	0	0	1	1	0
1	0	1	0	1	1
1	0	1	1	1	1
1	1	0	0	1	1
1	1	0	1	1	1
1	1	1	0	1	1
1	1	1	1	1	1

However the technique described suffers from two major limitations. The first of these is that, for any practical system, the truth table becomes too large to handle conveniently—this might be overcome by segmenting the table in some manner. The second disadvantage is a more fundamental one and is concerned with the sequential behaviour of the system.

A well known phenomenon in nonlinear systems is that of limit-cycle oscillation. The discrete mapping technique just developed gives little indication of the presence of limit-cycle oscillations and the Boolean equations describing the system are not well suited to an investigation of dynamic behaviour. We require, therefore, a discrete algebraic description of the required system which allows an exact digital realization and also allows the dynamic behaviour of the system to be investigated. The algebraic structures most suited to this problem are Finite (or Galois) Fields.

3. Galois fields

A Galois field[1] is an algebraic structure defined on a finite set of elements. Two operations, which we shall call multiplication and addition, are defined on all pairs of elements of the field according to a set of axioms. Galois fields, having k elements, exist for all $k = p^q$ where p is prime and q is integer;

TABLE 3. Arithmetic tables for GF(3).

×	0	1	2	+	0	1	2
0	0	0	0	0	0	1	2
1	0	1	2	1	1	2	0
2	0	2	1	2	2	0	1

such a field is designated GF(p^q). An example of a Galois field is arithmetic modulo a prime number p as shown in Table 3 for $p=3$. One of the important features of Galois fields is that arithmetic operations are closed (i.e. addition or multiplication of two elements from the field produces a result which is also an element of the field). This is obviously of importance when considering the digital realization of field operations and should be compared with normal addition and multiplication of a finite set of integers which are not closed operations.

Since the digital realization of a system will usually be in terms of binary variables, it is convenient to work in terms of Galois fields GF(2^q) since these contain 2^q elements which may be represented by q binary digits. The arithmetic tables for GF(2^2) are shown in Table 4. It is important to realize that the numbers in Table 4 are merely labels for the field elements and, as such, do not necessarily correspond to the natural integers. A very readable account of Galois field theory and applications is contained in ref. 1.

4. System representation by Galois field algebra

The objective is to model digital signal processing systems by the general finite-state machine of the form:

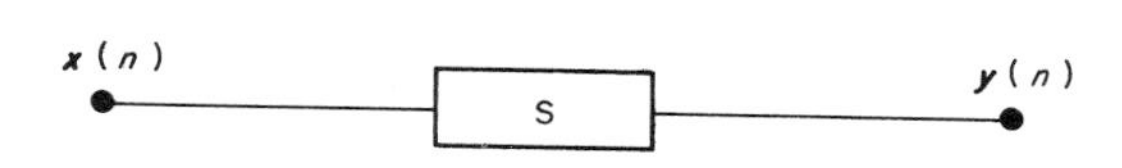

where $\boldsymbol{x}(n)$ and $\boldsymbol{y}(n)$ are respectively the discrete input and output vectors. In general the machine S has memory and therefore has a set of internal states $\boldsymbol{s}(n)$. The operation of the machine is to map from the set of states

TABLE 4. Arithmetic tables for GF(2^2).

×	0	1	2	3	+	0	1	2	3
0	0	0	0	0	0	0	1	2	3
1	0	1	2	3	1	1	0	3	2
2	0	2	3	1	2	2	3	0	1
3	0	3	1	2	3	3	2	1	0

and inputs into a set of outputs and new system states. This operation may be represented as:

$$\boldsymbol{s}(n+1)=f_1[\boldsymbol{s}(n)\boldsymbol{x}(n)] \tag{1}$$

$$\boldsymbol{y}(n+1)=f_2[\boldsymbol{s}(n)\boldsymbol{x}(n)] \tag{2}$$

This formulation is usually referred to as the Mealy model of a finite-state machine.

The problem to be considered is that of representing the functional relationships of equations (1) and (2) in terms of algebraic operations in a Galois field GF(p^q). In continuous algebra, the concept of representing general functional relationships in terms of power series or orthogonal functions is familiar; these concepts have a direct counterpart in finite algebra. A considerable amount of work has been reported[2] on the design of digital devices by means of orthogonal series and this work will not be described here.

4.1. Representation by finite polynomials

Before treating the general problem of representing the functions of equations (1) and (2) by means of finite polynomials over Galois fields, consider the autonomous (i.e. zero-input) behaviour of a particularly simple example of equation (1); one in which there is only a single state variable. Under these conditions, equation (1) reduces to

$$s(n+1)=f_1[s(n)] \tag{3}$$

$$s(n),\ s(n+1)\in \mathrm{GF}(p^q)$$

This functional relationship may be represented by a finite polynomial over a Galois field:

$$s(n+1)=\sum_{i=0}^{p^q-1} a_i s^i(n) \tag{4}$$

The polynomial coefficients may be determined from an exhaustive list of the values of $s(n)$ and the corresponding required values of $s(n+1)$. For example, in a system defined over GF(2^2), let the following list represent the behaviour of the system:

$s(n)$	$s(n+1)$
0	s_0
1	s_1
2	s_2
3	s_3

where the s_i are the required next-state values. Substituting these relationships in the polynomial, equation (4), gives the set of equations:

$$s_0 = a_0$$
$$s_1 = a_0 + 1a_1 + 1^2 a_2 + 1^3 a_3$$
$$s_2 = a_0 + 2a_1 + 2^2 a_2 + 2^3 a_3$$
$$s_3 = a_0 + 3a_1 + 3^2 a_2 + 3^3 a_3$$

Note that all arithmetic operations are those of $GF(2^2)$ (see Table 4). The set of equations may be written in matrix form as

$$\begin{bmatrix} s_0 \\ s_1 \\ s_2 \\ s_3 \end{bmatrix} = \begin{bmatrix} 1 & 0 & 0 & 0 \\ 1 & 1 & 1 & 1 \\ 1 & 2 & 3 & 1 \\ 1 & 3 & 2 & 1 \end{bmatrix} \begin{bmatrix} a_0 \\ a_1 \\ a_2 \\ a_3 \end{bmatrix} \qquad \text{or } \boldsymbol{s} = \boldsymbol{Ta} \tag{5}$$

It may be noted, in passing, that the section of the matrix enclosed in dotted lines is a number theoretic transform[3] over $GF(2^2)$. The matrix is non-singular and may be inverted to yield an expression for the polynomial coefficients in terms of the required state behaviour of the system:

$$\boldsymbol{a} = \boldsymbol{T}^{-1} \boldsymbol{s} \tag{6}$$

where

$$\boldsymbol{T}^{-1} = \begin{bmatrix} 1 & 0 & 0 & 0 \\ 0 & 1 & 3 & 2 \\ 0 & 1 & 2 & 3 \\ 1 & 1 & 1 & 1 \end{bmatrix}$$

A list of $\boldsymbol{T}$ and $\boldsymbol{T}^{-1}$ matrices is given in ref. 4.

A general system based on the polynomial representation will have the form:

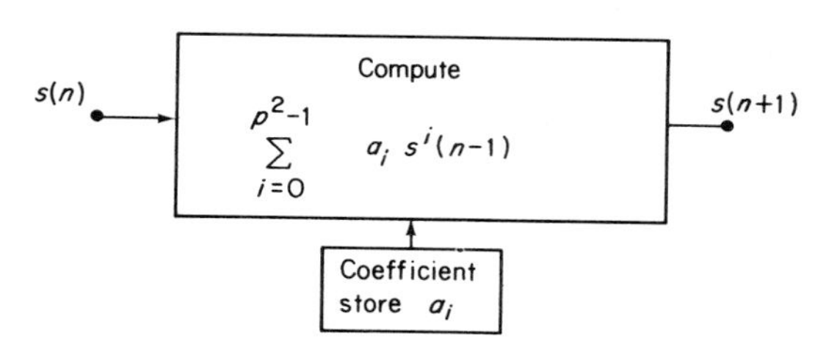

The polynomial representation has been demonstrated for only a single-variable system but may easily be extended to multi-variable systems defined by equations (1) and (2). For an r-variable system of the form

$$u=f(V_1, V_2, \ldots, V_r) \qquad u, V_1, V_2, \ldots, V_r \in \mathrm{GF}(p^q) \tag{7}$$

we may write

$$u=\sum_{i_1=0}^{p^q-1} \sum_{i_2=0}^{p^q-1} \cdots \sum_{i_r=0}^{p^q-1} a_{i_1 i_2 \ldots i_r} \prod_{j=1}^{r} V_j^{i_j} \tag{8}$$

A matrix expression may be obtained for the polynomial coefficients in a similar manner to the single-variable system previously studied.

The polynomial representation enables an arbitrary finite-state machine to be realized from the specified input–output state mappings. The original motivation for the use of Galois field theory was to study the dynamic behaviour of systems, and, in particular, to determine the class of systems which do not exhibit limit-cycle oscillations.

5. Stability of finite state machines

To illustrate the techniques for studying the stability of finite-state machines defined by polynomials over a Galois field, attention will be focused on the autonomous behaviour of a single state variable system:

$$s(n+1)=f[s(n)] \tag{9}$$

The state transition table associates a new state $s(n+1)$ with every possible existing state $s(n)$. For example, state transition tables in $\mathrm{GF}(2^3)$ might have the forms shown in Fig. 2.

The state behaviour shown in the first example represents an unstable system in that the state, and hence output, of the system can be one of two cycles of states. These cycles are precisely the limit-cycle oscillations which are observed in the normal realization of digital filters. The second example represents a stable system in that, regardless of the starting state, the system eventually reaches a ground state and remains there.

The problem to be considered is that of determining which subset of the set of possible polynomial coefficients gives rise to stable behaviour. The polynomial description of equation (9) is

$$s(n+1)=\sum_{i=0}^{p^q-1} a_i s^i(n) \qquad s(n),\ s(n+1),\quad a_i \in \mathrm{GF}(p^q) \tag{10}$$

Recursion of equation (9), or polynomial composition on equation (10), will yield an expression for the $(n+k)$th state in terms of the nth state. If

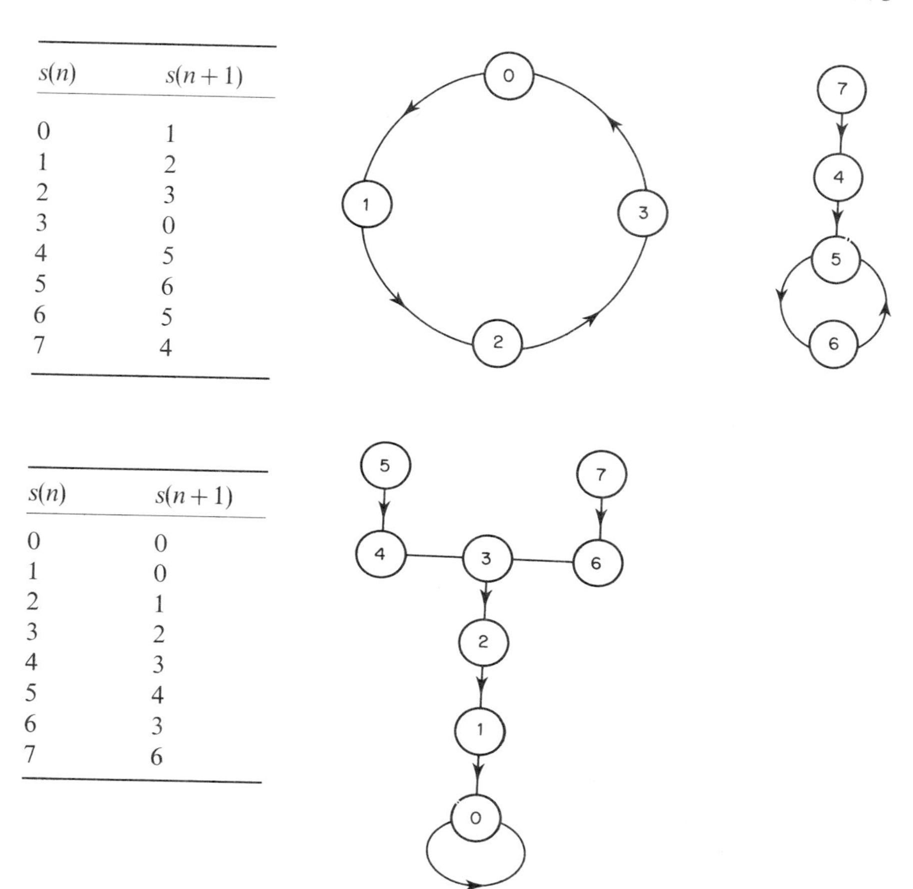

$s(n)$	$s(n+1)$
0	1
1	2
2	3
3	0
4	5
5	6
6	5
7	4

$s(n)$	$s(n+1)$
0	0
1	0
2	1
3	2
4	3
5	4
6	3
7	6

Fig. 2. State transition tables for machines over $GF(2^3)$.

$k = p^q - 1$ then, assuming that the required ground state is zero, the $(n + p^q - 1)$th state must be zero regardless of the value of the nth state. However this procedure is difficult to carry out. A method, whereby some of the results from linear finite state system theory may be applied, will now be described.

The polynomial of equation (10) may be expressed in vector form as

$$s(n+1) = \boldsymbol{a}'_1 \,.\, \boldsymbol{S}(n) \tag{11}$$

where:

$$\boldsymbol{a}'_1 = [a_{10} a_{11} \ldots a_{1(p^q-1)}] \quad \text{and} \quad \boldsymbol{S}'(n) = [1 \; s(n) \; s^2(n) \; \ldots \; s^{p-1}(n)]$$

Since the arithmetic operations in a Galois field are closed, polynomials may be written for powers of $s(n+1)$ in terms of powers of $s(n)$:

$$s^j(n+1) = \sum_{i=0}^{p^q-1} a_{ji}\, s^i(n) \tag{12}$$

where the a_{ji} are related to the set of coefficients a_{1i}. Therefore

$$s^j(n+1) = \boldsymbol{a}'_j \,.\, \boldsymbol{S}(n) \tag{13}$$

The complete set of equations may be written in matrix form as

$$\begin{bmatrix} 1 \\ s(n+1) \\ s^2(n+1) \\ \vdots \\ s^{N-1}(n+1) \end{bmatrix} \begin{bmatrix} 1 & 0 & 0 & \cdots & 0 \\ a_{10} & a_{11} & a_{12} & \cdots & a_{1(N-1)} \\ a_{20} & a_{21} & a_{22} & \cdots & a_{2(N-1)} \\ \vdots & & & & \vdots \\ a_{(N-1)0} & a_{(N-1)1} & a_{(N-1)2} & \cdots & a_{(N-1)(N-1)} \end{bmatrix} \begin{bmatrix} 1 \\ s(n) \\ s^2(n) \\ \vdots \\ s^{N-1}(n) \end{bmatrix} \tag{14}$$

where

$$N = p^q \qquad \text{or} \quad \boldsymbol{S}(n+1) = \boldsymbol{A}\boldsymbol{S}(n) \tag{15}$$

Equation (15) is a linear relationship but it must be remembered that all of the elements in the matrix $\boldsymbol{A}$ are functions of the polynomial coefficients a_i of equation (10).

The $(n+k)$th state may now be expressed as

$$\boldsymbol{S}(n+k) = \boldsymbol{A}^k\, \boldsymbol{S}(n) \tag{16}$$

Thus the study of the stability of nonlinear finite-state machines may be concentrated on the study of the properties of matrix $\boldsymbol{A}$. Many results concerning the stability and cycle structure of linear sequential systems may be applied.

6. Conclusions

An approach to the design of digital signal processing systems in terms of finite arithmetic structures has been described. There are still many problems to be solved and it is not certain that the techniques described here will necessarily lead to a complete theory for the design of digital systems. Two outstanding problems exist and these must be solved before the theory can find practical application. The first of these problems is to find methods for reducing the number of nonzero polynomial coefficients required for any particular system—methods are known for achieving this but, as yet, there is no systematic reduction method equivalent to the techniques for minimizing binary logic circuits. The second major problem is to map directly from the specification domain (e.g. time or frequency) of the required system into the finite arithmetic domain in which the system is to be realized.

References

1. H. S. Stone, "Discrete Mathematical Structures" Science Research Associates Inc. (1973).
2. M. G. Karpovsky, "Finite Orthogonal Series in the Design of Digital Devices", Wiley, Israel Universities Press.
3. J. M. Pollard, "The Fast Fourier Transform in a Finite Field", *Math. Comput.*, **25**, pp. 365–374 (April 1971).
4. D. H. Green, "Modular Representation of Multiple-Valued Logic Systems", *Proc. IEEE*, **121**, No. 6, pp. 409–418 (June 1974).

Evaluation of Two-dimensional Quantization

SABURO TAZAKI and YOSHIO YAMADA

Department of Electronic Engineering, Ehime University, Matsuyama, Japan

and

ROBERT M. GRAY

Department of Electrical Engineering, Stanford University, Stanford, U.S.A.

1. Introduction

One-dimensional quantization is the most commonly employed analog-to-digital technique. Gish and Pierce[1] and Davisson[2] show that for independent identically distributed (IID) random variables, optimum one-dimensional quantization results in an output entropy rate only 0·255 bits larger than the rate-distortion bound with a Gaussian signal and a squared-error fidelity criterion in the limit of large rate.

Several authors, Schützenberger[3], Amari[4], Zador[5], and more recently, Gersho[6] and Yamada and Tazaki[7], have developed asymptotic results for multi-dimensional quantization and have shown that strict improvement (lower entropy rate for a fixed average distortion or vice versa) over one-dimensional quantization is possible. In fact, it follows from rate-distortion theory[8] that, as the dimension tends to infinity, performance arbitrarily close to the rate-distortion bound can be achieved.

Multi-dimensional quantization means mapping a block of variables produced by a time series or multi-dimensional vector space such as colour television signals into another block of variables. The goal is to choose the encoded blocks so as to minimize the resulting average distortion under some constraint, for example fixed output entropy. We refer to this block-to-block quantization as vector quantization.

Here we develop the asymptotically optimum performance attainable with two-dimensional vector quantization on an IID Gaussian source.

2. Partition and entropy

Let a K-dimensional vector be $x=(x_1, x_2, \ldots, x_K)$, the joint probability density of x be $p(x)=p(x_1, x_2, \ldots, x_K)$, and the vector space of x be X. The performance of vector quantization depends on the choice of a partition of the space X into N disjoint subsets $S_1, S_2, \ldots, S_N$. For the simplicity of the discussion, we assume here that all subsets are bounded. The quantizer output vector $y=(y_1, y_2, \ldots, y_K)$ is determined by the membership in each $S_i (i=1, 2, \ldots, N)$. Thus we denote the quantizer output vector selected from the membership of S_i by $y(S_i)$.

It is well known that, in the one-dimensional space, uniform quantization is the most efficient one for large N in the sense of minimizing output entropy subject to a fixed average distortion.[1] In multi-dimensional quantization, each vector x in X should be approximated by the nearest neighbour vector $y(S_i)$.

Define the fidelity criterion by the distortion measure.

$$d_r(x, y)=\frac{1}{K}\sum_{k=1}^{K}|x_k-y_k|^r \tag{1}$$

where r is a positive integer and d_2 is the squared-error fidelity criterion. The average distortion between x and y is then

$$D_r=\sum_{i=1}^{N}\iiint_{S_i} d_r(x, y(S_i))p(x)\mathrm{d}x \tag{2}$$

According to Amari,[3] Zador[4] and Gersho,[6] the optimum two-dimensional vector quantization can be attained when the space X is partitioned by the regular identical hexagons. This partition is shown in Fig. 1, where each dot corresponds to the vector $y(m, n)=[y_1(m), y_2(m, n)]$. As shown in the figure:

$$y_1(m)=m\Delta/2 \qquad y_2(m, n)=\sqrt{3}(m+2n)\Delta/2 \tag{3}$$

where $m, n=0, \pm 1, \pm 2, \ldots$ and Δ is a quantizing unit.

The probability of the quantizer output vector $y(m, n)$ can be approximated by

$$Q(m, n)=\frac{\sqrt{3}\Delta^2}{2}\, p(y(m, n)) \tag{4}$$

provided that the probability density $p(x)$ varies smoothly in each hexagonal area. The constant factor $\sqrt{3}\Delta^2/2$ means the area of the unit hexagon. It follows from Equation (4) that the entropy per component of the vector y is

$$H \cong -\frac{1}{2}\sum_{m=-\infty}^{\infty}\sum_{n=-\infty}^{\infty} Q(m, n)\log Q(m, n) \tag{5}$$

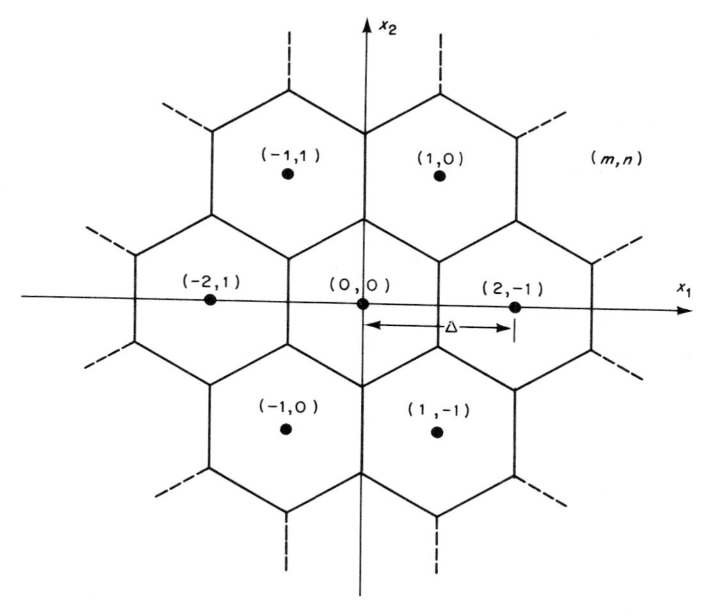

Fig. 1.

3. Gaussian source

Here, the squared-error fidelity criterion is used. The source density is assumed without loss of generality to be

$$p(x_1, x_2) = \frac{1}{2\pi\sigma^2} \exp\left[-(x_1^2 + x_2^2)/2\sigma^2\right] \tag{6}$$

where σ^2 is the variance of the source signal.

Substituting equation (6) into equations (4) yields

$$Q(m, n) = \frac{\sqrt{3}\delta^2}{4\pi} \exp\left[-\delta^2(m^2 + 3mn + 3n^2)/2\right] \tag{7}$$

where $\delta = \Delta/\sigma$. Denote by α the ratio D_2/Δ^2: this ratio tends to 5/72 as $1/\delta$ approaches infinity (see Appendix).

Substituting equation (7) into equation (5), we obtain

$$\begin{aligned} H &\cong \frac{1}{2} \log \frac{\sigma^2}{D_2} + \frac{1}{2} \log \frac{4\pi e\alpha}{\sqrt{3}} \\ &= \frac{1}{2} \log \frac{\sigma^2}{D_2} + 0{\cdot}227 \text{ (bits)} \end{aligned} \tag{8}$$

The first term of this equation represents the rate-distortion bound and the second term of it represents the quantization loss in entropy rate. This is 11% less than that of one-dimensional quantization.

4. Conclusion

The minimum quantization loss in entropy rate is definitively obtained for two-dimensional vector quantization of an IID Gaussian signal. This characterizes the strict improvement obtainable using the two-dimensional quantization instead of the one-dimensional quantization on the memoryless source considered.

Acknowledgment

This work was partially supported by the JSEP Program at Stanford.

References

1. H. Gish and J. N. Pierce, "Asymptotically Efficient Quantizing", *IEEE Trans. Information Theory*, **IT-14**, pp. 676–683 (Sept. 1968).
2. L. D. Davisson, "Rate Distortion Theory and Application", *Proc. IEEE*, **60**, pp. 800–808 (July 1972).
3. M. P. Schützenberger, "On the Quantization of Finite Dimensional Message", *Information and Control*, **1**, pp. 153–158 (1958).
4. S. Amari, "Theory of Information Space-Quantization and Bandwidth Compression of Signal Space", *J. IECE of Japan*, **48**, pp. 1709–1717 (Oct. 1965).
5. P. Zador, "Topics in the Asymptotic Quantization of Continuous Random Vectors", *Bell Telephone Lab. Tech. Memorandum* (Feb. 1966).
6. A. Gersho, "Asymptotically Optimum Block Quantization", 1977 ISIT at Cornell University, N.Y. (Oct. 1977).
7. Y. Yamada and S. Tazaki, "Vector Quantizing Method", 1977 Joint Conv. Record of the Societies Related with Electric at Shikoku District in Japan (Oct. 1977).
8. T. Berger, "*Rate Distortion Theory*", Englewood Cliffs, N. J., Prentice-Hall (1971).

Appendix

Since the source density $p(x)$ varies smoothly, the average distortion D_r can be approximated from equation (2) as follows:

$$D_r \cong \sum_{i=1}^{N} p(y(S_i)) \iiint_{S_i} d_r(x, y(S_i))\mathrm{d}x$$

$$\cong \frac{\iiint_{S_i} d_r(x, y(S_i))\mathrm{d}x}{\iiint_{S_i} \mathrm{d}x}. \tag{A.1}$$

Since the squared-error fidelity criterion is assumed, we obtain

$$D_2 \cong \frac{8}{\sqrt{3}\Delta^2} \int_0^{\Delta/2} \int_0^{(\Delta - x_1)/\sqrt{3}} d_2(x, z) \mathrm{d}x_2 \, \mathrm{d}x_1$$

$$= \frac{5\Delta^2}{72}, \tag{A.2}$$

where $z = (0, 0)$.

Consequently, we obtain $\alpha = 5/72$ as σ/Δ approaches infinity.

Topological Considerations in the Implementation of a Digital Filter using Microprocessors

JOSE PAULO BRAFMAN

Programa de Engenharia Electrica, COPPE-UFRJ, Rio de Janeiro, Brazil

JACQUES SZCZUPAK

Centro de Pesquisas de Energia Electrica, Cidade Universitária—Ilha do Fundao, Rio de Janeiro, Brazil

and

SANJIT K. MITRA

Department of Electrical and Computer Engineering, University of California, Santa Barbara, U.S.A.

1. Introduction

In recent years, due to the development of new digital techniques, digital filters are being increasingly used in many fields of application. Reduction in cost, size, and the higher speed of modern digital machines has made digital filters an attractive option in the design of a signal processor. However, these new digital machines have shorter wordlengths than the more costly general purpose digital computer. The purpose of this paper is to examine the possibility of using microprocessors[1] for the implementation of a digital structure to reduce the hardware complexity. These devices are being produced with lower costs, although with increasing complexity. As a matter of fact, technological achievements move so quickly that this fact must be incorporated into any realistic design philosophy. Otherwise, the final design will become obsolete in a fairly short time.

In general, microprocessors are designed in families[1] such that every new generation member is piecewise compatible with its predecessor. The new generation microprocessor has, however, a larger set of intrinsic functions and frequently a reduced instruction period. An example of the microprocessor evolution is the Intel 8 bit series that was initiated with an 8008 unit. This device has 48 different intrinsic functions and an instruction cycle of 12·5 μs. Its successor, the 8080 microprocessor, has an expanded set of 72 different intrinsic functions and an instruction cycle of 2 μs. Even faster devices using ECL technology are being reported.

The 8 bit characteristic of the 8000 family is not a rule,[3] but in general, reduced wordlengths correspond to lower costs and higher speeds for a microsystem. In addition to the wordlength restriction, absence of hardware arithmetic facilities must also be taken into account. The arithmetic operations, such as multiplication, are implemented in software form and, as a result, require a considerably long time. These restrictions are gradually being overcome one at a time by the rapidly developing microelectronics technology. It is expected that most of these problems will be solved in future generations if micro-processors. According to this reasoning, the implementation approach to be followed here is to enlarge the microsystem with modular hardwired external operators that are to be removed once the technology permits the removal.[2]

2. Digital filter representation and problems

The basic components in a digital filter structure are delays, adders, and multipliers. The digital network is obtained as an interconnection of these components such that no delay-free-loop[3,4] exists in the structure. Crochiere and Oppenheim[3] have shown that under this condition, it is always possible to obtain a description of the filter in a matrix equation in the form

$$\hat{W}(n) = \hat{F}\hat{W}(m) + \hat{G}\hat{W}(n-1) + \hat{H}X(n) \tag{1}$$

where the elements of $\hat{W}(n)$ are the ordered internal variables of the filter at time nT, corresponding to the nodes of the structure. $\hat{F}$ is a real lower triangular matrix with a diagonal of zeros, $\hat{G}$ and $\hat{H}$ are real matrices, and $\hat{X}(n)$ is the vector of inputs at the time nT. Due to the particular form of $\hat{F}$, between two sample clock pulses the internal variables are evaluated from the top to the bottom of $\hat{W}(n)$ as a function of known values only. Once $\hat{W}(n)$ is evaluated, it replaces $\hat{W}(n-1)$ in the next iteration.

In this work, the representation of (1) is used with a slight modification, by including the input vector $X(n)$ as one of the ordered variables. As a result, (1) is modified to the form

$$W(n) = FW(n) + GW(n-1) \tag{2}$$

where $W(n)$ is the new vector of ordered variables including the inputs as components. This change does not alter the evaluation procedure previously described.

The matrices F and G in equation (2) provide the necessary information concerning the topology and element values in the network. This, as well as the basis of operation defined on the elements in equation (2), is related to memory and speed requirements, and to the quantization problems. The influence of digital network topology is discussed next.

3. Topological limitations

Given a digital filter structure, Crochiere and Oppenheim[3] have shown the possible solutions by parallelism in order to increase the speed of operation of the filter. This approach is based on simultaneous operations in the network. If a microprocessor-based system is considered, the objective of this parallel type of operation is to divide tasks among members of a group of microprocessors in a multiprocessor type of operation. It is thus of practical interest to develop a procedure to divide the tasks and to determine a criterion for most efficient operation of the system. To this end, the critical path problem[5] is considered next.

3.1. The critical path problem

Consider a flow graph representation of a digital filter structure with all the delays removed.[4] The delay-free time-weighted flow graph is constructed from this subgraph by assigning a weight to each transmission branch, representing the time consumed in completing the corresponding operation. Additional weighted transmission branches are included in series with each input and output branch of the graph. These extra branches take care of the time spent to bring the variable into the subnetwork from an output or a delay component. A path that has the largest sum of weights (largest processing time) is called a *critical path*.[5]

Clearly, the critical path indicates a basic limitation in the digital filter operation, so that the minimum allowable delay between two consecutive sample clock pulses corresponds to this largest processing time. Other microprocessors may process the noncritical paths while the critical one is being executed.

To derive the time scheduling operation, it is necessary to include special transmissions in the delay-free time-weighted flow graph model to be used. For example, two paths intersect at a two-input summing node (or at a two-output pickoff node). In actual implementation, these paths may be processed by different microprocessors so that a variable must be transferred from one to the other. The time of transmission must be incorporated in the time-weighted flow graph by the inclusion of two special transmission

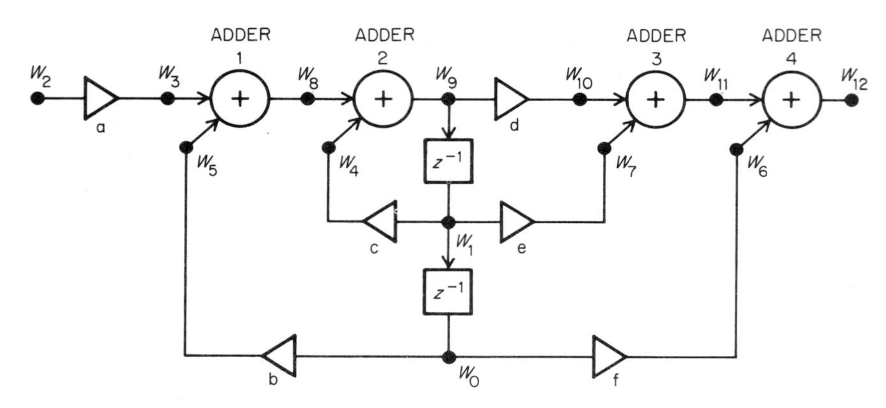

Fig. 1. A direct form structure.

branches with the corresponding variable transfer time. One of them is the output transfer time, the other is the input transfer time.

As an example, the direct form structure represented in Fig. 1 is analysed next. Here, the internal variables $\{w_i\}$ and the two-point adders are explicitly shown in the diagram. For this network, the exact form of equation (2) in the z-domain is

$$\begin{bmatrix} w_0 \\ w_1 \\ w_2 \\ w_3 \\ w_4 \\ w_5 \\ w_6 \\ w_7 \\ w_8 \\ w_9 \\ w_{10} \\ w_{11} \\ w_{12} \end{bmatrix} = \overset{\begin{matrix} 0 & 1 & 2 & 3 & 4 & 5 & 6 & 7 & 8 & 9 & 10 & 11 & 12 & \leftarrow\text{nodes} \end{matrix}}{\begin{bmatrix} 0 & z^{-1} & 0 & 0 & 0 & 0 & 0 & 0 & 0 & 0 & 0 & 0 & 0 \\ 0 & 0 & 0 & 0 & 0 & 0 & 0 & 0 & 0 & z^{-1} & 0 & 0 & 0 \\ 0 & 0 & 1 & 0 & 0 & 0 & 0 & 0 & 0 & 0 & 0 & 0 & 0 \\ 0 & 0 & a & 0 & 0 & 0 & 0 & 0 & 0 & 0 & 0 & 0 & 0 \\ 0 & c & 0 & 0 & 0 & 0 & 0 & 0 & 0 & 0 & 0 & 0 & 0 \\ b & 0 & 0 & 0 & 0 & 0 & 0 & 0 & 0 & 0 & 0 & 0 & 0 \\ f & 0 & 0 & 0 & 0 & 0 & 0 & 0 & 0 & 0 & 0 & 0 & 0 \\ 0 & e & 0 & 0 & 0 & 0 & 0 & 0 & 0 & 0 & 0 & 0 & 0 \\ 0 & 0 & 0 & 1 & 0 & 1 & 0 & 0 & 0 & 0 & 0 & 0 & 0 \\ 0 & 0 & 0 & 0 & 1 & 0 & 0 & 0 & 1 & 0 & 0 & 0 & 0 \\ 0 & 0 & 0 & 0 & 0 & 0 & 0 & 0 & 0 & d & 0 & 0 & 0 \\ 0 & 0 & 0 & 0 & 0 & 0 & 0 & 1 & 0 & 0 & 1 & 0 & 0 \\ 0 & 0 & 0 & 0 & 0 & 0 & 1 & 0 & 0 & 0 & 0 & 1 & 0 \end{bmatrix}} \begin{bmatrix} w_0 \\ w_1 \\ w_2 \\ w_3 \\ w_4 \\ w_5 \\ w_6 \\ w_7 \\ w_8 \\ w_9 \\ w_{10} \\ w_{11} \\ w_{12} \end{bmatrix}$$

where z^{-1} is the delay operator.

For the implementation of this filter, the particular weights to be assumed for the operations, without any loss of generality of the method, are: 2 u.t. (units of time) for the transfer of an input word to the microprocessor memory or for a variable in this memory to be transferred to the output of the filter or to another position in the memory (a delay operation), 6 u.t. for floating-point multiplication including the transfers of variables both ways between memory and a special hardwired multiplier, and 12 u.t. for floating-point addition, also including all the time necessary for the transfer of variables from and to microprocessor memory. These time values are related to a special hardwired operator expansion of an 8080 system being built. The minimum time necessary for a single microprocessor implementation of the structure of Fig. 1 is then 92 u.t., which takes into account input-output transfers, ($2 \times 2 = 4$ u.t.), multiplication ($6 \times 6 = 36$ u.t.), addition ($4 \times 12 = 48$ u.t.), and delay transfers ($2 \times 2 = 4$ u.t.).

If several microprocessors are available to implement this network, the time required to transfer variables from one to the other must also be included in the delay-free time-weighted flow graph. Without any loss of generality, these transfer times are assumed to be equal to 2 u.t. Figure 2 represents the time-weighted flow graph version of the network of Fig. 1. In the flow graph, all variable transfer times are indicated by oriented hexagons. The multipliers, adders, and internal variables in Fig. 1 are repeated in Fig. 2 for convenience. However, each input of the adders in this last figure has been weighted (shown by a hexagon) to include the variable transfer time. One of these corresponds to the time required to transfer a variable from a microprocessor to an output register. The other hexagon represents the time required to transfer the variable from the output register into a second microprocessor. Up to this point, no additional notation has been used to represent the input or the output transfer time separately. For instance, each of the variables w_3 and w_5 belongs to a different path in the time-weighted flow graph of Fig. 2. Both paths intersect at node 15. Each

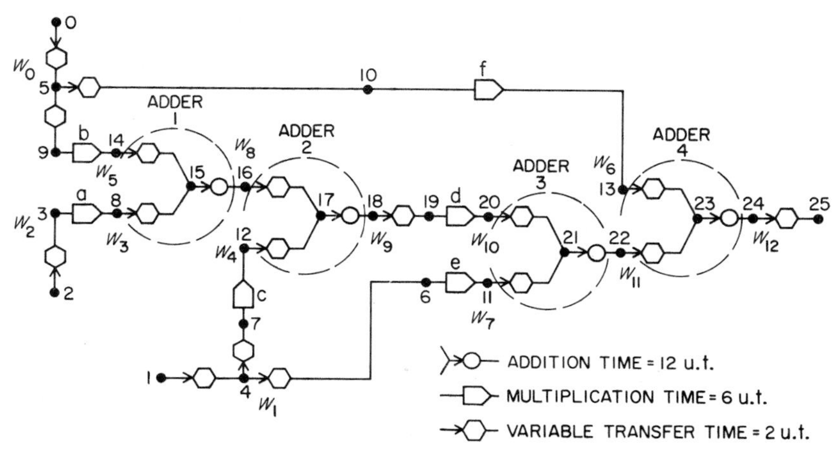

Fig. 2. A time-weighted signal flow graph representation of the network of Fig. 1.

path may be executed by a different microprocessor, or w_3 may be transferred to the microprocessor where w_5 is located or vice versa. In any case, one of the paths must be discontinued at node 15. The choice between the possible paths is discussed in the time scheduling procedure.

It is important to observe in Fig. 2 that each input to the flow graph is either an input to the network or an output of a delay component. In any case, hexagons in series with each of these inputs take care of the variable transfer time. For example, the variable corresponding to node 0 will generate the variable w_0 only after 2 u.t.

This last variable belongs to more than one path (pickoff node). Similar to the case of the two-input adder, a hexagon is located in series with each of the paths outgoing the pickoff node. One of them is the time required to transfer w_0 to the one or more output registers (one in this case). The other hexagons (one of each path initiating at node 5) correspond to the time required to bring the variable from one of the output registers into the proper microprocessor. Again, the basis to choose which hexagon corresponds to what type of transfer is discussed in the time scheduling procedure.

Another case of interest here is concerned with the variable w_9 (node 18). This variable is used as an input to the multiplier with coefficient d. However, it also generates the variable at node 1. As this last variable might be located in a second microprocessor, a hexagon is inserted between nodes 18 and 19 to take care of this possibility.

Clearly, in some cases, it may not be necessary to transfer variables from one microprocessor to another. Consequently, the corresponding hexagons should be removed; otherwise, the total processing time will be more than is needed. This is possible once the paths are determined by the time scheduling procedure.

3.2. Time scheduling procedure

First, the connection matrix[5,6] C_1 is defined, associated with G_1, the delay-free time-weighted subgraph. C_1 is a square matrix, relating the delay-free time-weighted flow graph variables, ordered according to the precedence relation[3] and such that the (i, j)th element $c_1(i, j)$ is either: (1) zero if there is no transmission from node i to node j, or (2) equal to the time of operation indicated in the transmission from node i to node j. Therefore, C_1 is an upper triangular matrix with a diagonal of zeros.

Any path in G_1 can be followed by proper analysis of C_1. For instance, if a path begins at node i and $c_1(i, j) \neq 0$, then it will go to node j. It will proceed from node j to a node k if $c_1(j, k) \neq 0$. The accumulated time spent from node i to node k is given by $c_1(i, j)+c_1(j, k)$. This operation can be best performed by using the lower part of C_1 and the precedence relation. The time scheduling algorithm is as follows.

Step 1: Location of the critical path

(a) Starting from the first row of the connection matrix, for every nonzero $c_1(i, j)$, $j>i$, make $c_1(j, i)=c_1(i, j)+k_i$, where k_i is the largest member of the set, $\{c_1(i, l), l<i\}$.

(b) The largest element in each row to the left of the main diagonal is located. The row that has the largest of all these elements indicates the node where the critical path ends. This largest number equals the total number of u.t. needed to process the critical path.

(c) To determine the critical path, the row R and column C of this node are stored. Next, the largest element in row C is located and its position stored. The operation is repeated, tracing back to the initial node, identified by a row of zeros to the left of the main diagonal.

Step 2: Removal of the critical path

Eliminate the critical path from the graph. If a nonempty graph is left, return to Step 1 to determine the next critical path; if not go to Step 3.

Step 3: Distribution of tasks

The minimum processing time necessary to operate the filter is obtained by letting one microprocessor handle the critical path and assigning the other paths to other micros, while preserving the precedence relation.

The flow graph in Fig. 2 illustrates the algorithm. The upper triangular part of matrix A_1 corresponds to the upper triangular part of the connection matrix C_1. In the lower triangular part of A_1 the result obtained after application of Step 1(a) is indicated on the next page. No zero elements are shown in A, beside those in the main diagonal, to improve clarity.

Applying Step 1(b) to A_1, it follows that the critical path $CP1$ has a minimum processing time of 76 u.t. and it terminates at node 25. According to Step 1(c), the critical path is traced back. It is then represented by the sequence of nodes together with the corresponding accumulated processing time as shown below:

$$CP1=0\text{–}5\text{–}9\text{–}14\text{–}15\text{–}16\text{–}17\text{–}18\text{–}19\text{–}20\text{–}21\text{–}22\text{–}23\text{–}24\text{–}25$$

$$\text{Cumulative time} \quad 0 \; 2 \; 4 \; 10 \; 12 \; 24 \; 26 \; 38 \; 40 \; 46 \; 48 \; 60 \; 62 \; 74 \; 76 \qquad (5)$$

The next step, Step 2, is the removal of $CP1$ from the subgraph and the

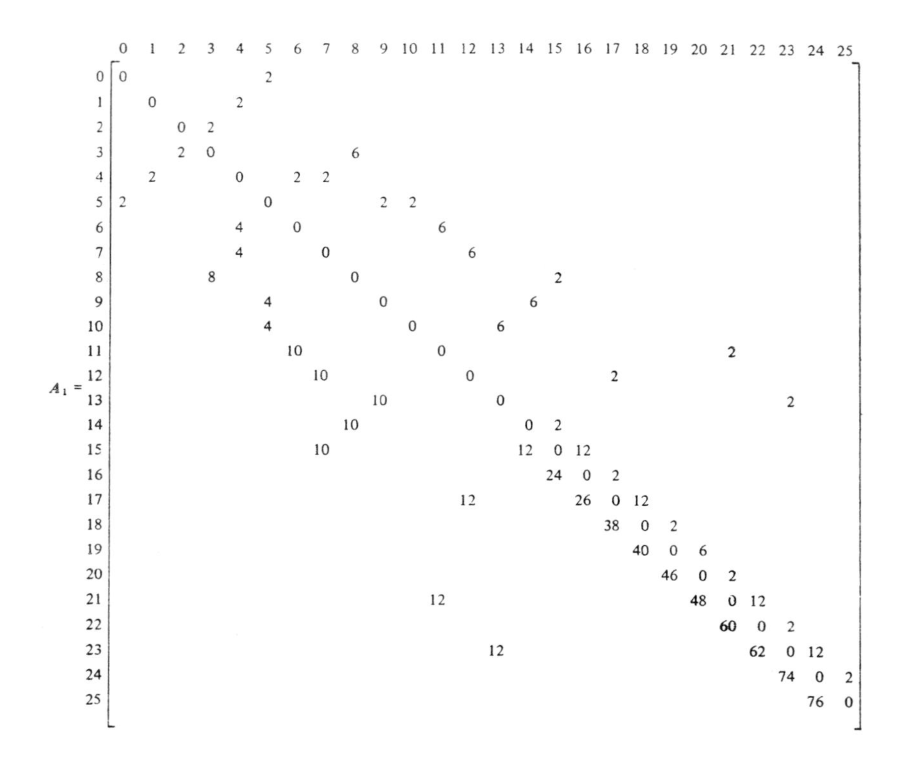

beginning of a new cycle. After several cycles of application of Steps 1 and 2, collection of paths is obtained together with the corresponding processing times:

$$CP2 = 1\text{–}4\text{–}7\text{–}12\text{–}17 \qquad \text{(processing time} = 12\text{ u.t.)}$$
$$CP3 = 4\text{–}6\text{–}11\text{–}21 \qquad \text{(processing time} = 10\text{ u.t.)}$$
$$CP4 = 5\text{–}10\text{–}13\text{–}23 \qquad \text{(processing time} = 10\text{ u.t.)}$$
$$CP5 = 2\text{–}3\text{–}8\text{–}15 \qquad \text{(processing time} = 10\text{ u.t.)} \qquad (6)$$

After removing $CP5$, the remaining subgraph is empty. The algorithm is now at Step 3, where all the critical paths are to be distributed with respect to $CP1$. The objective here is to add a minimum processing period separating two consecutive sample clock pulses. This is shown in Fig. 3, where $CP2$, $CP3$, $CP4$, and $CP5$ are distributed so as to match the $CP1$ requirement while preserving the sequence of operations. As an illustration, observe that after passing by node 8, there is a 2 u.t. delay due to the transfer of variable to the input of the adder. Therefore, only after 10 u.t. the

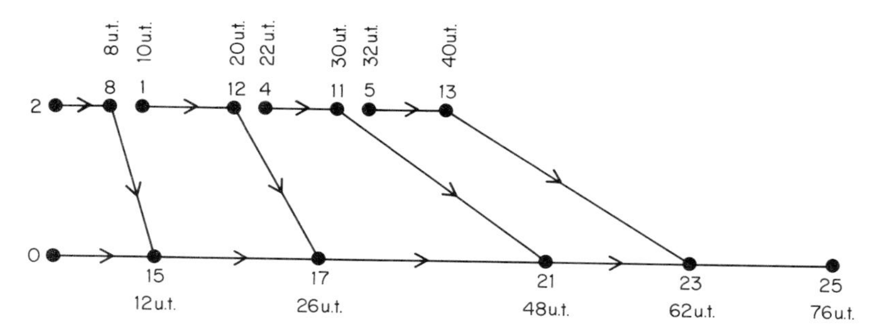

Fig. 3. Two microprocessor distributions of tasks.

second input, node 1, is initiated. The other values in Fig. 3 are evaluated accordingly.

Clearly, the addition of another microprocessor allows the realization with a processing time of 76 u.t. The small reduction with respect to the 92 u.t. of the single processor operation is due to the reduced amount of parallelism in the structure. This first topological constraint is added to the need of incorporating extra delays for multiprocessor operation, motivated by the exchange of variables.

It is important to notice that there is a certain degree of flexibility when constructing the time-weighted signal flow graph. This is because of the order in which the two-input adders are selected to realize an adder with more than two inputs. This ordering is a designer's choice. However, as the accumulated time is known for all the input nodes of the adder, a choice corresponding to a best order is always possible. In the example, nodes 8, 12, and 14 are evaluated with processing times 8 u.t., 10 u.t., and 10 u.t., respectively. Therefore, changing the order of additions to be performed will not reduce the critical time. Only the critical path will have changed.

3.3. Ordering of variables and memory

Consider a single path delay-free network. In such a type of structure only one variable needs to be stored at a time to evaluate the output variable. This is illustrated in Fig. 4, where $w_1, w_2, \ldots, w_{n-1}, w_n$ are evaluated in the

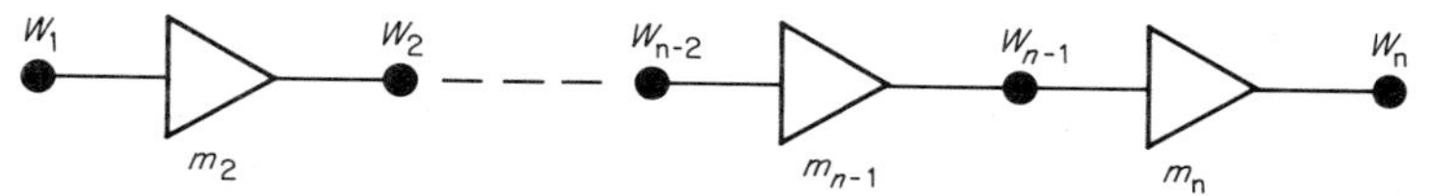

Fig. 4. A single-path delay-free network.

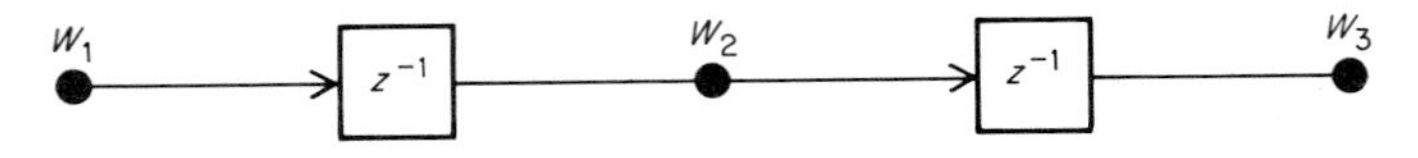

Fig. 5. A delay-only network.

same order and may be stored in the same memory location. This is possible whenever the previously evaluated variable is not necessary for further use.

An extension of this single path property is to assign one memory location to each of the critical paths. In this case, disregarding exchange of variables from one micro to the other and the delay component requirements, the total amount of memory is sufficient for the processing needs.

To reduce the number of memory locations, the inverse path sequence is determined in the delay-only subnetwork. This structure consists only of the interconnected delay components of the filter. The inverse path sequence is indicated for the subnetwork in Fig. 5 as $w_3 - w_2 - w_1$. Here, $w_3(n)$ is to be evaluated first, as $w_2(n-1)$. Next $w_2(n)$ is evaluated as $w_1(n-1)$. If the precedence relation is to be followed, the previous values of variables would require extra memory locations. Other memory requirements are imposed by the mode of operation used in the microprocessor system.

4. Conclusion

This work proposes simple solutions for a microprocessor based digital filter implementation. The approach proposed takes advantage of the newly arrived microprocessor generations, while compensating temporary limitations by the use of external hardwired operations. Following this approach, a digital filter system is being built using the Intel 8080 microprocessor as the basic CPU. To reduce the operating time between sample clock pulses, external hardwired operations providing faster floating-point multiplication and addition facilities have been incorporated to the system. The time units mentioned in the time scheduling operation refer to this particular system. The details of this general purpose digital filter unit are available elsewhere.[7]

In addition, topological limitations were discussed, together with a time scheduling algorithm to improve speed and memory requirements by employing a multiprocessor type of operation.

References

1. Editorial Group, "Microprocessors", *Electronics*, Special Issue on Microprocessors (April 15, 1976).
2. A. C. Davies and Y. T. Fung, "Interfacing a hardware multiplier to a general purpose microprocessor," *Microprocessors*, **1**, pp. 425–432 (Oct. 1977).

3. R. E. Crochiere and A. V. Oppenheim, "Analysis of linear digital networks," *Proc. IEEE*, **63**, pp. 581–595 (April 1975).
4. J. Szczupak and S. K. Mitra, "Detection, location, and removal of delay-free loops in digital filter configurations," *IEEE Trans. Acoustics Speech and Signal Processing*, **ASSP-23**, pp. 558–562, (Dec. 1975).
5. S. Seshu and M. B. Reed, "Linear Graphs and Electrical Networks". Reading, MA: Addison-Wesley, 1961.
6. M. J. Lilenbaum, "Modelo Pert-Cpm-Sistematica de Sua Aplicacão a Administracão e Projetos". Rio de Janeiro, R. J., Brasgraf (Mar. 1970).
7. J. P. Brafman, "Filtros digitais usando microprocessandores," M.Sc. Thesis, COPPE/UFRJ, Rio de Janeiro, Brazil (July 1977).

Part 4
APPLICATIONS

4.1. Applications to speech processing and communications

Variable Rate Speech Processing†

B. GOLD

Massachusetts Institute of Technology, Lincoln Laboratory, Lexington, U.S.A.

1. Introduction

Present day digital speech processors are capable of transforming an analog voice signal into a bit stream at a great variety of rates. For example, the standard rate for digital voice over the public telephone system is considered to be 64 kbps (kilobits per second). On the other hand, channel vocoders operating at 1200 bps (bits per second) can, in many cases, provide a reasonably satisfactory voice channel for conversational speech. Given this wide range of rates it is natural to inquire as to the potential advantages of a multiple rate digital speech processing algorithms. This paper addresses some of the issues raised by such an inquiry and describes, in some detail, a particular such scheme that has been implemented on a high-speed programmable processor.

The two advantages associated with multiple rate speech processing are: (a) bit rates can be lowered in response to time-varying network requirements; and (b) bit rates can be raised to improve the speech quality, especially under degraded environmental conditions. In case (a) this means, for example, that the network will respond to heavy traffic loads by causing the average throughput rate between speech terminals to decrease thus avoiding the necessity of terminating existing connections or generating many busy signals. In case (b), for example, a user in an airplane where the heavy acoustic noise background creates the need for a higher rate speech processing algorithm may be able to request a steady high rate by network priority methods. The precise way that a network design could be well

---

† The views and conclusions in this paper are those of the contractor and should not be interpreted as necessarily the official policies, either expressed or implied, of the United States Government.

matched to a multiple-rate speech terminal is a subject for continuing research and development and cannot be discussed in great detail in this paper. However, as a preliminary to the main content of this paper, it is useful to discuss, in broad terms, some overall network flow control strategies that would apply.

2. Network considerations

Perhaps the most straightforward method of flow control could occur at dial-in. As the connection is being made, the network assesses its capability and "advises" the requesting terminal to operate at no more than some prescribed rate. Thus, as users hang up and new users enter, the overall network throughput can be adjusted in response to the time-varying network traffic statistics. Depending on the number of terminals, this flow control procedure may cause total flow to change too slowly (for example, if many new users request connections while the "old" users are still on) and could lead to the need for busy signals.

Another method could be denoted as exclusive user control. Here the user can manually raise or lower his transmitted rate depending on his own assessment of the situation plus discussion with the receiving user. For example, a new user might begin operating at 2400 bps and then raise his rate if the receiving user told him that his speech was garbled. Or he might be informed that the high rate at which he was sending was in fact not being received whereupon he could lower his rate. Or the user might decide that he is not in an acoustically benign environment and should begin at the highest possible rate for improved robustness.

An extension of this strategy would be a more general feedback strategy wherein both the end users as well as the network exchange flow information and adjust rates accordingly. Such schemes can lead to unstable network situations and must be carefully controlled.

Finally, a strategy of "embedded coding" or network bit-stripping can be used. In this case, the user terminal tends to be sending at a high rate but the network node, if it becomes too congested, strips away part of this information before relaying it to the next node or to the end user. This strategy has the advantage but there is no feedback and hence no danger of instability. However, it puts severe constraints on the design of the speech processing terminal since there are few speech algorithms that can survive bit-stripping. In the work reported here, emphasis has been placed on variable rate methods that would work in an embedded coding environment.

3. The multi-rate channel vocoder

The idea behind the proposed multi-rate system is based on a merging of two speech processing methods; one, the channel vocoder[1] and two, the sub-band coder.[2] The channel vocoder, shown in Fig. 1, consists of a bank of filters for spectrum analysis, another filter band for synthesis and a pitch and voicing detector to control the excitations. The sub-band coder shown in Fig. 2 consists of a filter bank, a sampler and quantizer bank for analysis and a decoder bank for reconstruction of the received sampled, quantized information and a final filter bank. Imagine that the banks of bandpass filters in Figs. 1 and 2 were identical. Imagine also that the quantizer in Fig. 2 was a one-bit quantizer. Then Fig. 3 would be a reasonable way of performing sub-band coding; in essence, the one-bit signal from the analysis bandpass filter would be modulated by the average energy emanating from that filter. But Fig. 3 now closely resembles a channel vocoder, with the difference being only in the excitation signals. This immediately suggests the multi-rate channel vocoder scheme shown in Fig. 4, where *both* pitch-derived excitation and sub-band excitation are transmitted and, *for each and every channel*, the network and the receiver choose which of the two excitation signals to use. Since sub-band excitation does not depend on a pitch measurement, it should yield a more robust system; thus, as the rate is increased by adding sub-band channels, system performance would be expected to improve.

4. Spectral flattening

Preliminary experiments with the system indicated that the robustness under acoustic background noise environments is clearly improved by adding sub-band excitation. However, this excitation introduced an unpleasant roughness to the output speech. This roughness can be explained by examining the distortion introduced by the sampling and quantizing. Figure 5 shows the frequency response curves of the analysis filter bank. A bandpass, or integer band[2] sampling technique was applied; that is, for each channel the system sampling rate of 8264 Hz was divided by an integer to yield a lower rate f_s such that $\frac{1}{2}nf_s$ and $\frac{1}{2}(n+f)f_s$ were respectively the lower and upper edges of the filter pass-band. Since the filter parameters were based on an existing channel vocoder design,[3] integer band sampling was not optimally efficient with respect to bit rate, but this fact did not perturb the fundamentals of the experiment. Figure 6 indicates the nature of the distortion introduced by the one-bit quantization superposed on integer band sampling. Assume that a sinusoid at 1100 Hz is present in a filter

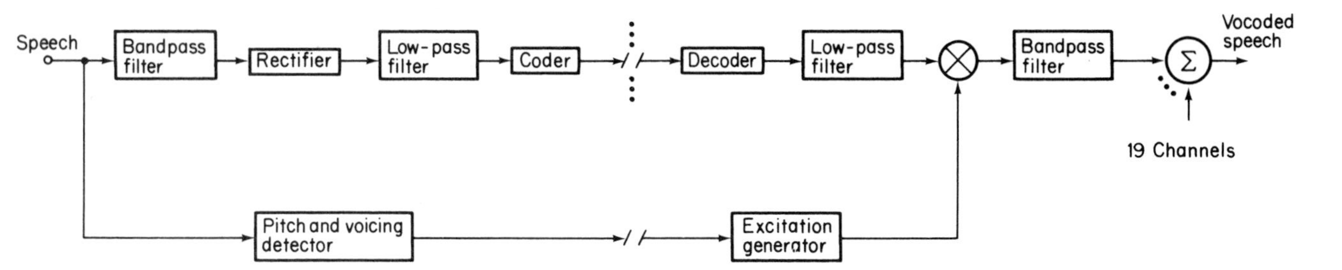

Fig. 1. Channel vocoder.

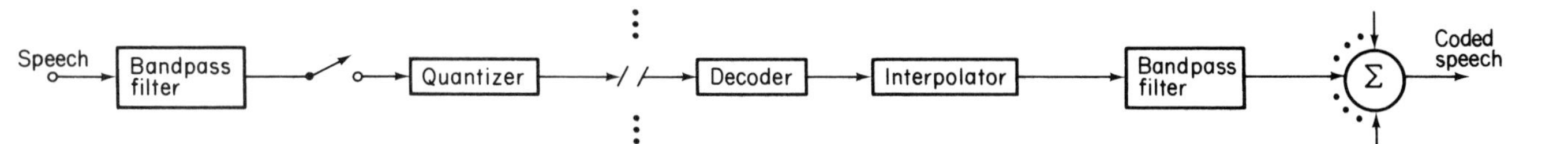

Fig. 2. Sub-band coder.

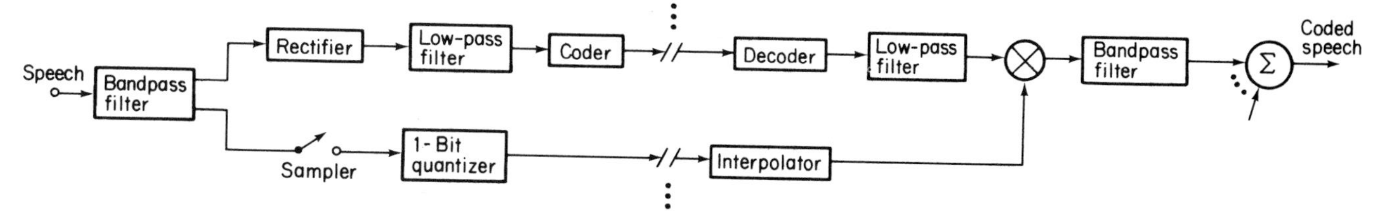

Fig. 3. Sub-band coder with one-bit quantizer.

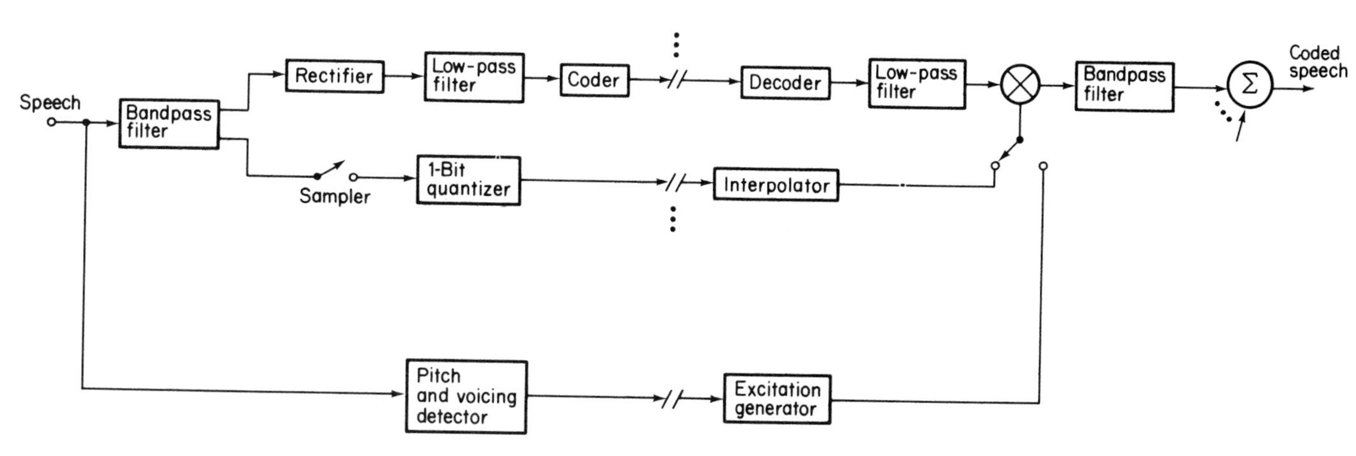

Fig. 4. Multirate vocoder with optional sub-band coding.

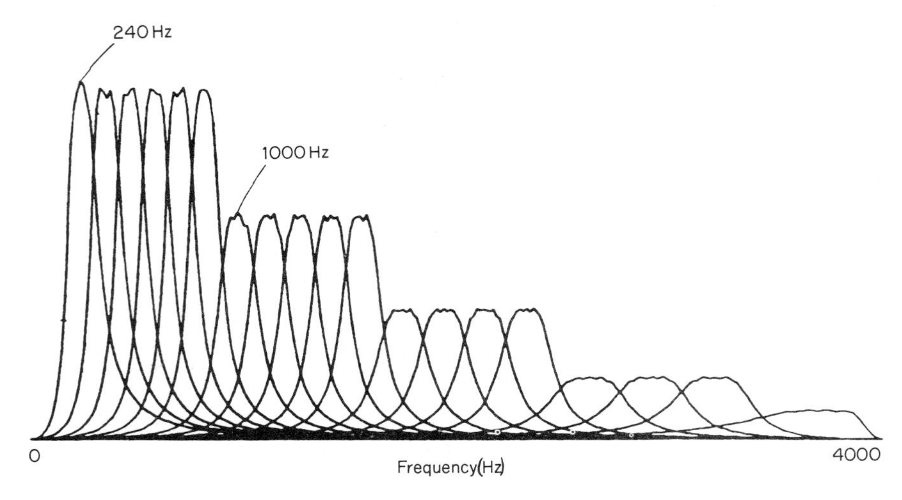

Fig. 5. Belgard filter bank (analyser).

centred at 1000 Hz. The sampling process creates the periodic spectrum indicated by the dotted repetition of the filter response. The one-bit quantizer creates a square wave with frequency components at 3300 Hz, 5500 Hz, etc. These distortion products fold back into the original filter and thus remain within that filter band even after interpolation. Since noise within a narrow band is better masked by a strong signal within the same band, it was hoped that the perceptual effect of these distortions would not prove too annoying, but such was not the case.

The roughness was considerably attenuated by introducing a bandpass filter and band limiter in the sub-band excitation path. It is well known from noise theory[4] that a bandpass limiter will increase signal-to-noise ratio if the signal is initially above the noise. Thus, the multi-rate system of Fig. 7 proved to be a great improvement over that of Fig. 4. Introducing these additions in every channel results in a system that is a spectrally flattened channel vocoder[5] with multi-rate sub-band excitation. This is the system we listened to extensively. In addition to improving the quality when sub-band excitation was used, spectral flattening also smooths out fluctuations caused by large pitch variation, pitch errors and buzz-hiss transitions.

5. Simulation facilities

Most of the listening was conducted with a nonreal-time but high-speed processing system, as shown in Fig. 8. Speech was stored on a disc and then

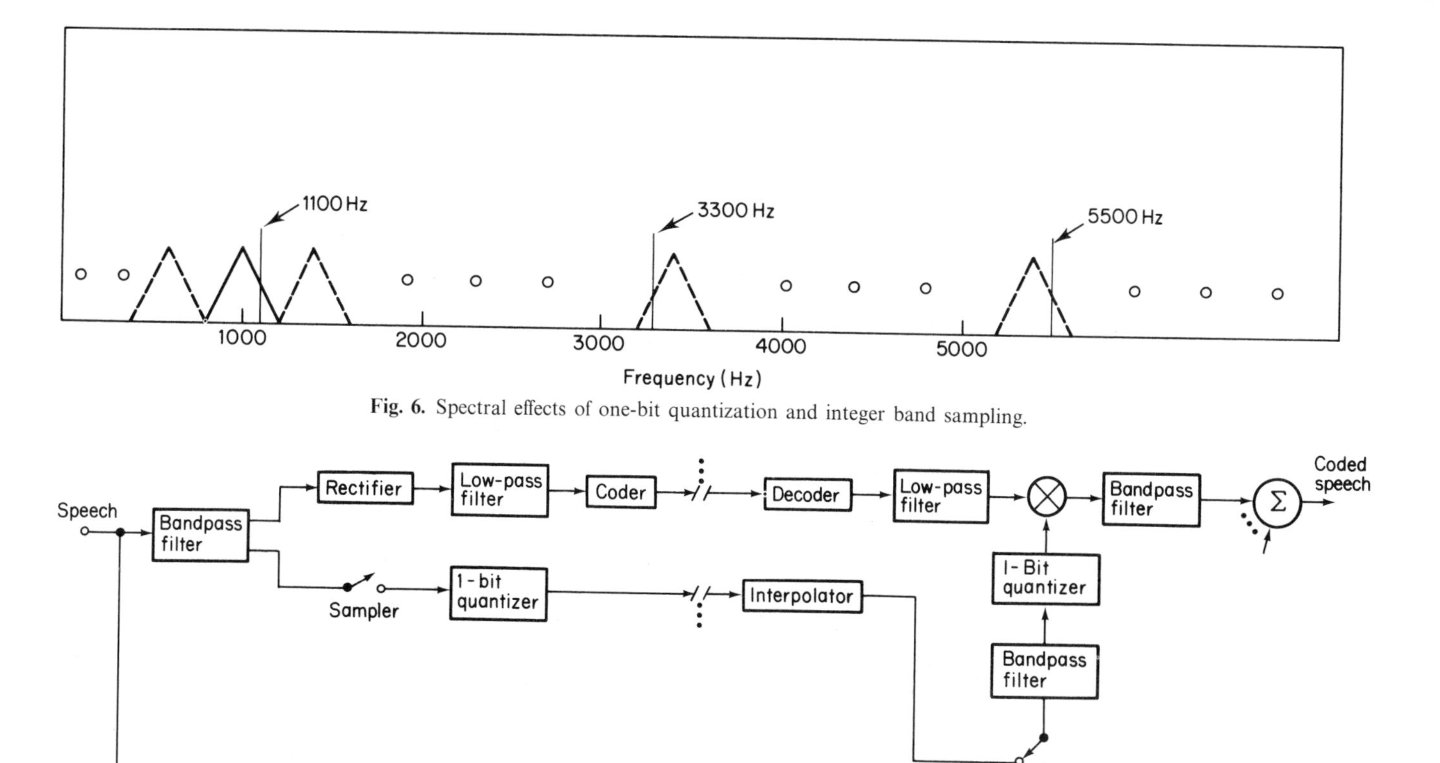

Fig. 6. Spectral effects of one-bit quantization and integer band sampling.

Fig. 7. Multirate spectrally flattened vocoder with optional sub-band coding.

processed on a batch basis through the LDSP (Lincoln Digital Signal Processor), a sequential processor with a 50 nanosecond cycle time and 100 nanosecond multiplier. Typically, processing time was about thrice real time. Thus, it was feasible to indulge in extensive listening with a large variety of parametric conditions and for many different speech samples.

More recently, a real-time version of the system has been implemented using two LDSPs, one for analysis and one for synthesis.

6. Informal listening

In listening to this system, we looked for three things. First we wanted to verify that increased sub-band excitation yielded greater robustness. Second, we wanted to know which channels were best to excite. Finally, we wanted to compare the overall performance of this system with other systems (such as adaptive predictive coding (APC)) at comparable rates.

It was easily verified that sub-band excitation increased robustness. We played tapes containing airborne command post noise, fighter cockpit noise, telephone speech and even music and in every case, the addition of 3 to 9 channels of sub-band excitation increased both the intelligibility and quality of the synthesized speech; the greater the number of sub-band excitation channels, the greater the improvement.

It is interesting to note that for clean speech, the results were not as clearcut. The system, when run as a spectrally flattened 2400 bps channel vocoder, generally produced speech of excellent quality. Furthermore, the sub-band excitation still introduced some slight hoarseness so that for many utterances it would be difficult to choose the better quality result. This, however, may be advantageous in a network environment if we expect the rates to switch rapidly. If the speech sounded too different at different rates, then the switching could be expected to introduce very annoying subjective

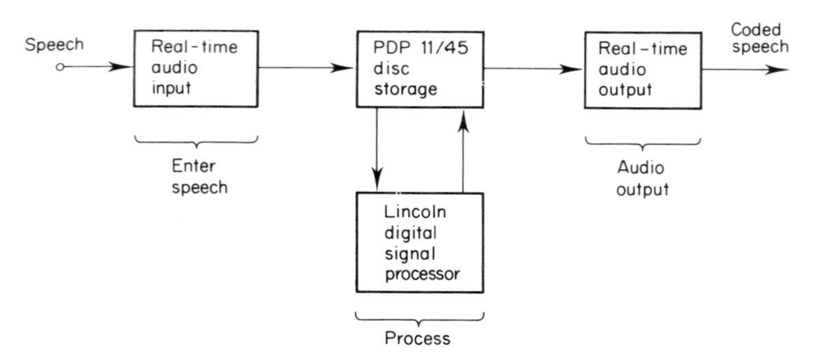

Fig. 8. Three steps in nonreal-time facility.

effects and would probably make the system unusable. This was definitely *not* a problem with our system.

For the digital filter bank we used, wherein each bandpass filter was a four-pole butterworth filter designed via bilinear transform[1], it was ascertained that sampling rates had to be of the order of 1–2 kbps per channel. Since we preferred to maintain a rate not to exceed 16 kbps, this usually led to a choice of 4–6 channels for the sub-band excitation, the remaining bands being excited via pitch and voicing information. We tried many different combinations of channel numbers; since the variety of possibilities is very large, it is difficult to arrive at anything like an optimum. The following general observation appears valid; when sub-band excitation was applied to higher frequency channels, roughness was less perceptible but robustness was impaired; the reverse was true when sub-band excitation was applied to lower frequency channels. As a compromise, most of our production runs used channels 3 through 8 for sub-band excitation; this corresponded to filters with centre frequencies of 480, 600, 720, 840, 1000, 1150 Hz.

Finally, we listened to a comparison of our system with APC.[6] To compare with 8 kbps APC we excited channels 1, 2 and 3 of the vocoder with the sub-band signals which yielded about 8 kbps. Our source tapes consisted of (a) paragraphs read by both male and female speakers with a good quality microphone in a quiet environment; (b) an excerpt from an Edward R. Murrow broadcast that included some background music; (c) an acoustic recording of Bessie Smith singing "St. Louis Blues" and sentences read by a male speaker in a severe airborne command post (ABCP) noise environment. The general impressions of several experienced listeners were as follows:

(a) For the male and female paragraphs, the sub-band system was preferred.
(b) For Murrow, the APC reproduced the music background somewhat better but on an overall basis, the sub-band system was preferable.
(c) For Bessie Smith, APC was superior.
(d) For the severe ABCP noise background, the tendency was to favour the sub-band system.

We then compared 16 kbps APC[7] with the vocoder with channels 1 through 6 excited by sub-band, which yielded about 16 kbps. Here the situation was reversed, APC being generally preferred for all cases, this despite the fact that the 6 channel sub-band excitation was clearly more robust than the 3 channel sub-band excitation.

7. Discussion and summary

Simulation of a spectrally flattened channel vocoder with added sub-band excitation has demonstrated a straightforward design of a multi-rate speech terminal that is matched to a network with embedded coding capability. With this structure a large number of rates are theoretically possible. Furthermore, rapid switching of rates does not appear to disturb listeners' perception of speech quality. Since the integer band sampling was applied to an existing filter bank, the sampling rates we obtained were appreciably higher than the rates that could have been obtained if the filter bank were designed to efficiently accommodate integer band sampling.

The system suffered from several shortcomings. For some of the utterances, a roughness was introduced that listeners felt was unpleasant. This was most noticeable when the input speech was of high quality and appeared to be aggravated for higher pitched voices. For speech with a high acoustic noise background, addition of the sub-band excitation smoothed out the extreme roughness caused by pitch errors but intelligibility was still quite impaired, presumably because of spectral errors introduced by the processing of the noisy background.

Although the system studied is approximately twice as consumptive of computational power as the more conventional single filter bank vocoder such as Belgard,[3] recent studies by Blankenship[8] indicate that the application of CCDs to the design should yield an implementation that will be quite compact.

Acknowledgment

This work was sponsored by the Defense Advanced Research Projects Agency.

References

1. Rabiner and Gold, "Theory and Applications of Digital Signal Processing," Chapter 12, Englewood Cliffs, NJ: Prentice-Hall, 1975.
2. Crochiere, Webber and Flanagan, "Digital Coding of Speech in Sub-bands," *BSTJ*, **55**, No. 8 (Oct. 1967).
3. D. M. Dear, Private Communication (1966).
4. Davenport and Root, "Random Signals and Noise." New York. McGraw-Hill, 1958.
5. Tierney, *et al.*, "Channel Vocoder with Digital Pitch Extractor," *J. Acoustical Society America* **36**, 1901–1905 (1964).
6. Atal and Schroeder, "Adaptive Predictive Coding of Speech Signals," *BSTJ*, **49**, pp. 1973–1986 (Oct. 1970).
7. Seneff, Private Communication.
8. Blankenship, Private Communication.

Selection of a PCM Coder for Digital Switching

K. SHENOI and B. P. AGRAWAL

ITT Telecommunications Technology Center, Stamford, U.S.A.

1. Introduction

Rapid advancement in large scale integration technology, the steep decline in digital-hardware cost, and ease of encryption, storage and processing have accelerated the transition from analog to digital transmission and switching. The change-over is accompanied by a trend towards increased digital processing of voice-frequency signals. When the signal enters and leaves a transmission and switching system via an analog transducer, e.g. a telephone set, the digital signal processing must be preceded and followed by analog-to-digital converter (ADC) and digital-to-analog converter (DAC), respectively. In traversing the chain analog–digital-transmission–analog the signal suffers degration because of aliasing and quantization, assuming impairement-free transmission.

There is yet another conversion which may introduce additional signal degradation. For digital signal processing, it is necessary to represent the signal, internal to the digital switch, in a linear PCM format. However, to interface with digital trunks, external to the digital switch, it is usually necessary to compress (convert) the linearly coded signal into 8-bit words according to either the μ-Law or A-Law companding schemes.[1] Subsequent to transmission, the compressed signal is expanded (converted) back into a linear PCM format.

The noise introduced by quantization and code conversion tends to degrade system performance which is measured in terms of signal-to-noise ratio (SNR), idle-channel noise (ICN), and dynamic range (DR). Some of the parameters which determine these measurements are: number of bits per linear PCM word, amount of d.c. offset, companding characteristic as specified by CCITT, and the probability density function of the signal. The dependence of SNR, ICN, and DR on the above parameters is presented below in a tutorial fashion. Specifically, we will answer such practical (often asked by designers) questions as: Does a 13-bit coder perform better than the μ-Law coder? Is the SNR obtained from 14-bit linear coder always 6 dB better than that obtained by 13-bit coder? Is the cascade of a 13-bit linear coder and μ-Law compressor compatible with a μ-Law expander? Will the cascade of a 14-bit uniform coder and A-Law compressor perform better than the cascade of a 13-bit uniform coder and A-Law compressor? The answers to three questions will be facilitated by a review of the input–output characteristics of various coders. The review is followed by analysis of SNR performance in Section 3 and of idle-channel behaviour in Section 4. The questions regarding compatibility with CCITT specifications are investigated in Section 5. The paper ends with a summary of important conclusions.

2. Input–output characteristics

The function of a PCM coder is to transform an analog signal, continuous in both time and amplitude, into a sequence of binary words. Equivalently, a coder assigns a code word, out of a finite set of words or symbols, to each pulse amplitude modulated (PAM), discrete-time and continuous amplitude, sample. The simplest commonly used coder is the uniform or "linear" coder. An n-bit (including sign) coder is defined by specifying a set of $N+1$ decision levels $s_0, s_1, \ldots, s_N$, a set of N output levels $x_1, x_2, \ldots, x_N$, and the relation $x=Q(s)$ where

$$Q(s)=x_i \quad \text{if} \quad s_i-1<s\leqq s_i \tag{1}$$

Observe that the input–output characteristic (IOC) has a staircase form as shown in Fig. 1. The end-point levels s_0 and s_N define the clipping, or overload, levels which are sometimes referred to as "virtual decision levels". In telephony, a mid-tread characteristic, shown in Fig. 1(a), is preferred over a mid-riser characteristic, shown in Fig. 1(b), for reasons of lower idle-channel noise (ICN). For mid-riser coders $N=2^n$ and for mid-tread coders $N=2^n-1$, where one level is sacrificed to obtain a symmetrical IOC and "dead zone" about the origin.

The noise introduced in coding (in fact, in quantizing) is modelled as an additive random component e:

$$e=Q(s)-s \tag{2}$$

and is approximated for large N as being independent of the signal. The quantization noise is termed granular noise if the input sample satisfies the inequalities $s_0\leqq s\leqq s_N$. On the other hand, if the input is not bound by s_0 and s_N, the quantization noise is described as overload noise. The designer

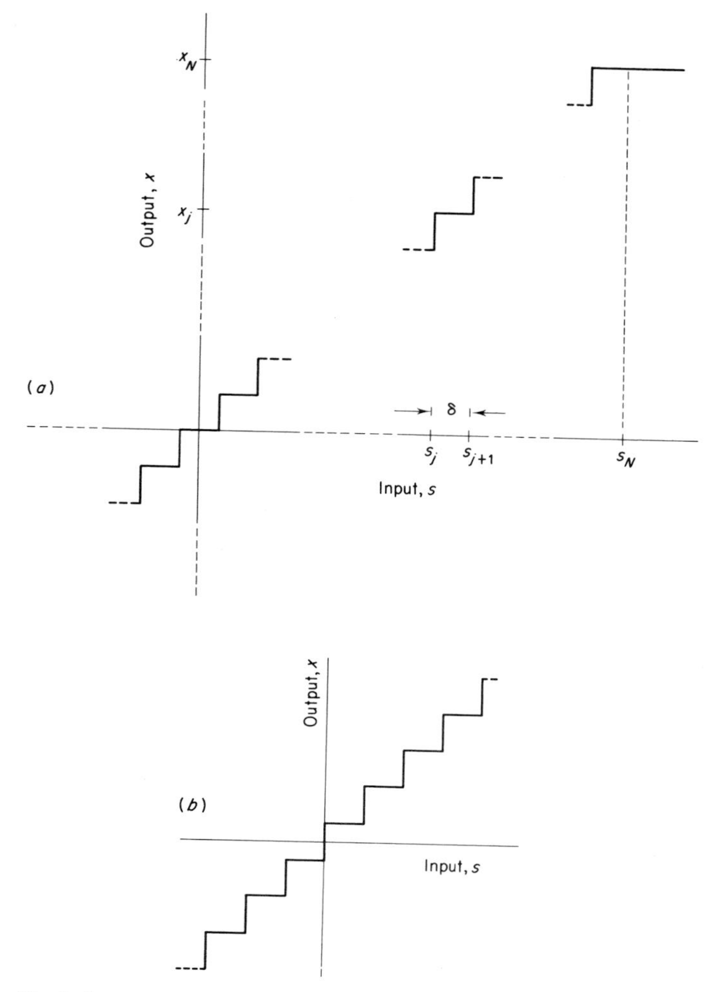

Fig. 1. Input–output characteristics of mid-tread and mid-riser coders.

can trade granular noise for overload noise, and vice-versa, by selecting the clipping levels appropriately.

When the decision levels are equally spaced so that the interval $\delta_i = s_i - s_{1-1}$ is not a function of i, the coder is described as a uniform coder and the IOC has steps of equal width and equal height. In telephony applications, where the Laplacian probability density function is known to describe adequately the distribution of the speech PAM samples, uniform

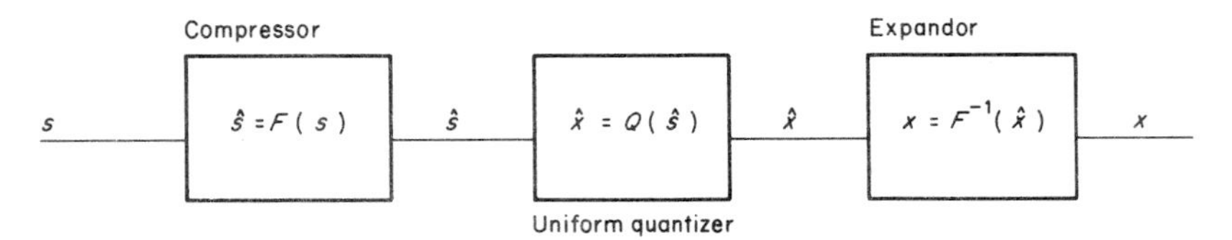

Fig. 2. Scheme for illustrating the operation of companding coders.

quantization is not the most effective scheme to achieve good performance. For a given number of quantization levels N, nonuniform spacing of the decision levels, optimized for probability density, will yield a higher signal-to-noise ratio and will make the SNR less sensitive to variations in the input signal power. The design of such nonuniform IOCs, that is, selecting the decision levels s_i and the output levels x_i, is described in ref. 2.

An effective technique for studying a nonuniform quantizer is to view it as a cascade of a "compressor" followed by a uniform quantizer. The effect of uniformly quantizing the compressed signal is to allocate more output levels in the range where signal amplitudes have high probability at the expense of making large errors in the range where signal amplitudes have a low probability of occurrence. The overall scheme depicted in Fig. 2 is called companding. The compressor function $F(s)$ is, roughly, the inverse of the cumulative probability density function of the input signal.

Speech signals can be characterized by the Laplacian probability density function which belongs to the exponential class. Hence, the optimal compressor would be a logarithmic one. In practice, the logarithm is replaced by piecewise linear approximations. The 15-segment μ-Law and 13-segment A-Law as defined by the CCITT, use two different approximations to the optimal compressor and their definitions serve primarily as industrial standards. An excellent description of companding coders is available in ref. 3.

3. SNR performance

The degradations suffered due to quantization and code conversion will, as usual, be measured in terms of signal-to-noise ratio. The input speech levels considered here will be greater than 60dBm0; the behaviour of coders for input signals below this level is the subject of Section 4. It will be assumed that:

(a) The coders are symmetric.

(b) The 13-segment A-Law and 15-segment μ-Law codes will be those recommended by the CCITT.[1] The clipping level (the amplitude of a full load sinusoid) specified there is

1. 3·14 dBm0 for the A-Law.
2, 3·17 dBm0 for the μ-Law.

(c) The analog signal (speech) has a Laplacian probability density function.

(d) The coder will be preceded and followed by ideal low-pass filters to avoid aliasing.

(e) The 13- and 14-bit coders will be of the mid-tread type. A clipping level of 3 dBm0 will be assumed unless a code conversion is considered.

(f) Whenever a code conversion occurs the coders will be normalized to have the same clipping level.

From ref. 2 the error, quantization noise, is given by

$$e = x - s \tag{3}$$

and the noise power is given by

$$\overline{e^2} = E\{e^2\} = \int_{-\infty}^{\infty} (s-x)^2 \, p(s)\mathrm{d}s \tag{4}$$

where E is the expectation operator and $p(s)$ the probability density function of S. For speech signals:

$$p(s) = \frac{1}{\sqrt{2}\sigma} \exp\left(\frac{-\sqrt{2}|s|}{\sigma}\right)$$

(the Laplacian density) where σ^2 is the signal power. For an n-bit uniform coder with stepsize σ, e^2 can be expressed as

$$e^2 = \sum_{j=1}^{N} \int_{s_{j-1}}^{s_j} (x_j - s)^2 \, p(s)\mathrm{d}s + 2\int_{s_N}^{\infty} (s - x_N)^2 \, p(s)\mathrm{d}s \tag{6}$$

In the above formula, the symmetry property of the coder IOC as well as that of the probability density has been exploited. The second term represents the noise due to saturation (the overload noise).

In Fig. 3, curves of the SNR versus input signal power are shown for 14-bit and 13-bit uniform coders. It is seen that over quite a large range of input signal power, the 14-bit coder shows a 6 dB improvement over the 13-bit coder. However, over this range, the source of the noise is primarily the granularity of the quantizer. At high signal levels, both coders yield the *same* SNR since in this region the dominant contribution to the error arises from overload noise. SNR curves for the μ-Law and A-Law, taken from ref. 1, are shown in Fig. 4. Comparison of Fig. 3 and Fig. 4 demonstrates that the 13-bit coder outperforms both μ-Law and A-Law coders.

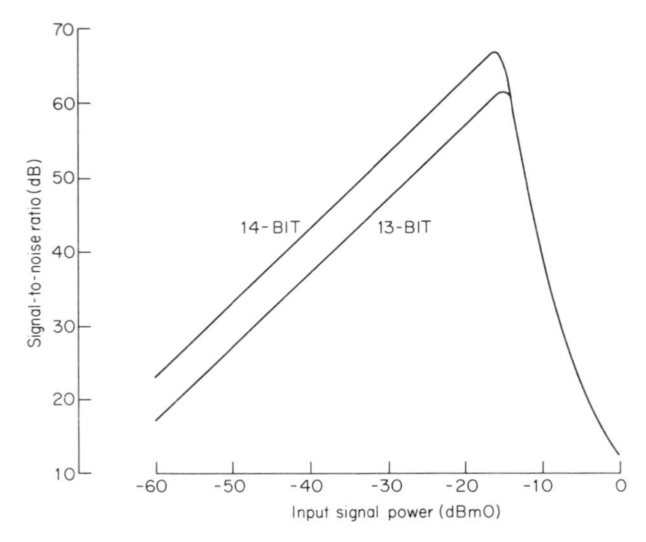

Fig. 3. Signal-to-noise ratio versus input power for 13- and 14-bit uniform coders.

The degradation caused by code conversion was analysed according to the schemes depicted in Fig. 5. The analysis also serves as a convenient way to compare the uniform and companding coders considered here. The conversion between the uniform and companded codes was characterized in

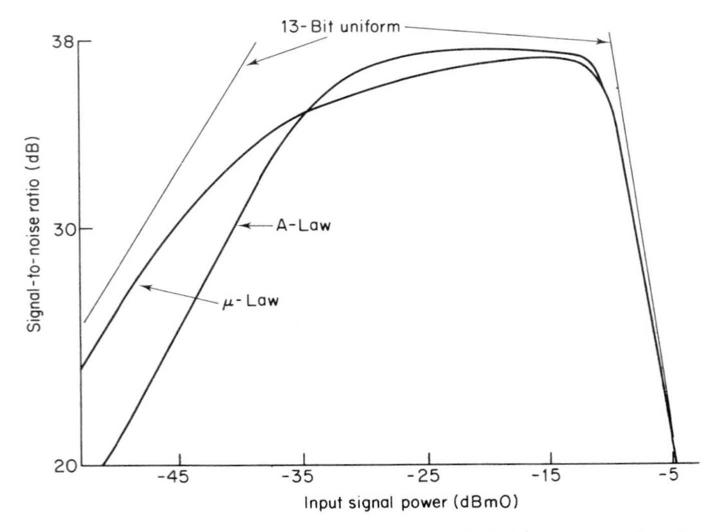

Fig. 4. Signal-to-noise ratio versus power for μ-Law and A-Law coders taken from ref. 1.

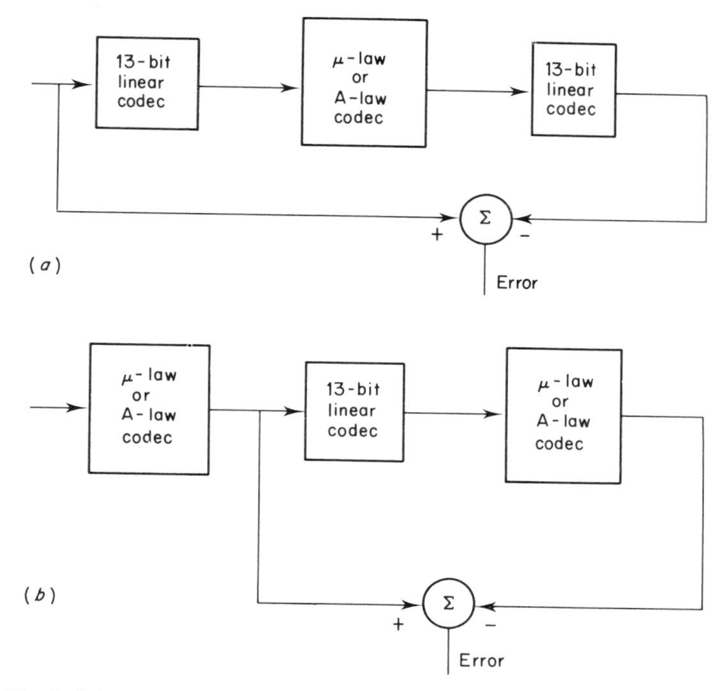

Fig. 5. Schemes for analysing the distortion introduced by code conversions.

the following way:

(a) The companded level Z, corresponding to a uniform code output value x is the value closest to

(i) $\dfrac{4096\ x}{2^{n-1}-0{\cdot}5}$ (A-Law)

(ii) $\dfrac{8159\ x}{2^{n-1}-0{\cdot}5}$ (μ-Law)

(b) The uniform PCM level y, corresponding to a companded code output level z, is obtained by rounding to n bits the quantity:

(i) $\dfrac{z(2^{n-1}-0{\cdot}5)}{4096}$ (A-Law)

(ii) $\dfrac{z(2^{n-1}-0{\cdot}5)}{8159}$ (μ-Law)

The SNR curves obtained for the configuration of Fig. 5(a) are shown in Fig. 6. An immediate inference is that, if code conversions are used, the

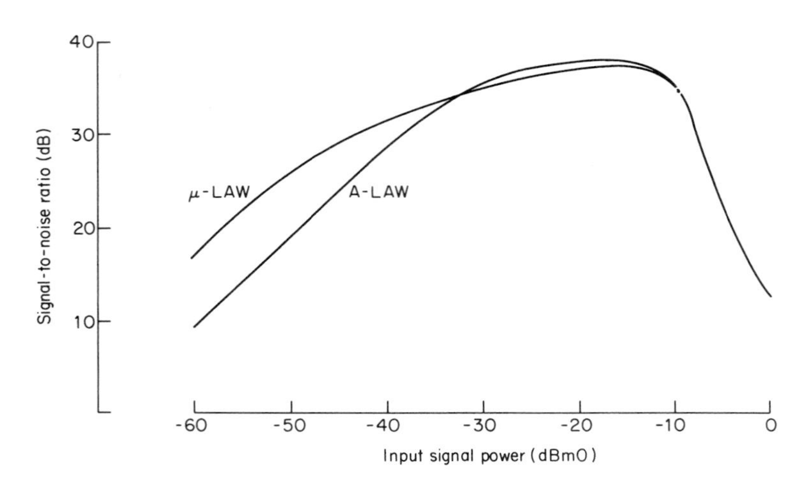

Fig. 6. SNR curves for the scheme in Fig. 5(*a*).

overall SNR is typically that of the companding coder. Using a 14-bit coder, instead of a 13-bit coder, in Fig. 5(*a*), would yield a small improvement in the overall SNR. This improvement was found to be less than 1·7 dB or 0·8 dB for A-Law or μ-Law intermediate codes respectively.

The conversion error defined in Fig. 5(*b*) was found to be zero. That is, for either of the uniform codes (13- or 14-bit) and either of the companding laws (A or μ), the companded code is faithfully reproduced at the output. The overall degradation introduced by the chain analog–companded–uniform–companded–analog would be that of the companding coder itself; the intermediate conversion to linear would not be apparent.

4. Idle-channel behaviour

When the speech signal is absent at input, the coder is said to be in an idle state. In this state, the input is a weak interference signal such as noise or crosstalk. Often the input noise is enhanced by the coder. The enhancement varies with the amount of interference and is noticeably pronounced in the presence of d.c. offset or bias. Under these circumstances, the low-pass filter following the decoder will output an undesirable signal, called idle-channel noise. In this section, the idle-state performance of both linear (13-bit and 14-bit) and companded (μ-Law and A-Law) coders is examined. The output noise power depends both on the bias (d.c. offset) and input noise power. Accordingly, the output noise power is calculated as a function of input noise power for various values of d.c. offset.

For the idle-state performance evaluation, the input interference signal will be modelled as a white Gaussian random process whose variance (power) is σ^2. The performance with or without d.c. offset will then correspond to using a non-zero mean ($m \neq 0$) *or zero* mean ($m=0$) random process, respectively.

The mean and mean-squared values of the coder output are given by

$$\bar{x}=\sum_j x_j \text{ Prob. } \{s_{j-1}<s\leqq s_j\} \tag{7}$$

and

$$\overline{x^2}=\sum_j x_j^2 \text{ Prob. } \{s_{j-1}<s\leqq s_j\} \tag{8}$$

respectively. The summation in both equations (7) and (8) are taken over all possible output levels (2^{n-1} for an n-bit linear code). The probabilities are calculated by integrating the Gaussian probability density function over the appropriate intervals. For example:

$$\text{Prob. } s_{j-1}<s\leqq s_j=\frac{1}{\sqrt{2\pi\sigma}}\int_{s_{j-1}}^{s_j} \exp\left[\frac{(s-m)^2}{2\sigma^2}\right]\mathrm{d}s \tag{9}$$

where m and σ^2 are respectively the mean and variance of the input signal s. The output noise power can be expressed in terms of x and σ^2 as follows:

$$\begin{aligned}\text{Output noise power} &= E\{(x-x)^2\}\\ &=\overline{x^2}-\bar{x}^2\end{aligned}$$

The removal of the d.c. component for calculating the power is a reflection of the fact that p- and c-message weightings[4] would assign a null weight to the d.c. component.

For clarity of presentation, the following discussion is divided into two broad categories. The first part deals with the behaviour of coders in the absence of a d.c. offset and the second in the presence of d.c. offset.

4.1. Without a d.c. offset

The dependence of output noise on input noise power in the range -60 dBm0 to -90 dBm0 is shown in Fig. 7. It was considered unlikely that, in the idle state, the input power level would exceed -60 dBm0. It is seen that as the input noise level is decreased, the output level for an A-Law coder remains constant at -66 dBm0. This occurs because the A-Law coder has a mid-riser IOC and, consequently, exhibits a saturation level. In contrast, the μ-Law and uniform coders have a mid-tread IOC and show a sharp drop in output noise level as σ^2 is reduced below a certain "cutoff" (the cutoff power is the input level below which the coder has an attenuation of greater than

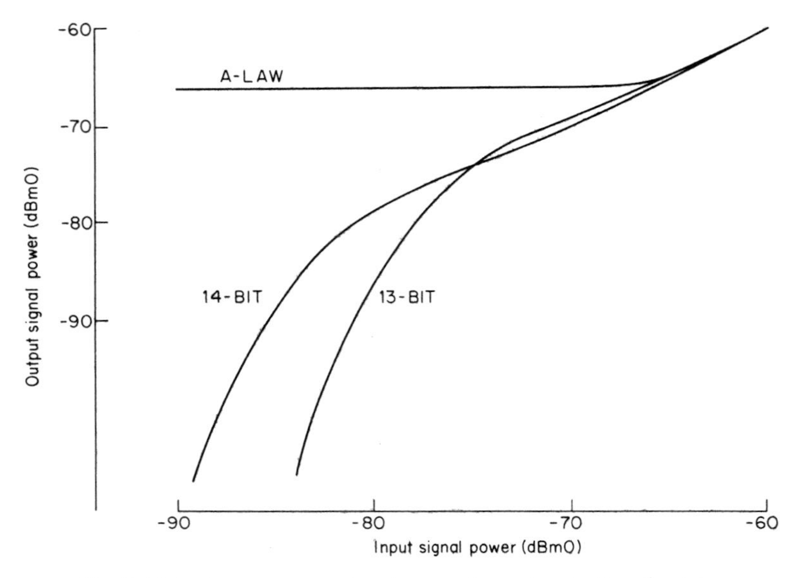

Fig. 7. Output power versus input power for coders in the idle state.

1 dB) which is about -78 dBm0 for 13-bit uniform and μ-Law, and -84 dBm0 for 14-bit uniform.

For certain input levels, it is seen that the output level is greater than the input. This enhancement, or amplification, is noticeable only when the noise power σ^2 lies in the range

$$\delta^2/16 \leqq \sigma^2 \leqq \delta^2/4$$

for Gaussian distributions. The maximum amplification was found to be about 1·25 dB for the μ-Law and uniform coders and occured at input noise levels just above the cutoff power. The worst-case amplification, for binary distributions of variance $0{\cdot}25\delta^2$, could be as much as 6 dB.

4.2. With a d.c. offset

A d.c. offset has an adverse effect on mid-tread coders in the sense that the cutoff power is reduced and the amplification increased. A. d.c. offset of half a stepsize would cause a mid-tread coder to behave like a mid-riser coder and exhibit the phenomenon of saturation. Figure 8 shows the effect of adding a half-step for the 13-bit and 14-bit linear coders. Both coders show a saturation value for the output noise power which is -72 dBm0 for the 13-bit coder, and -78 dBm0 for the 14-bit coder.

codecs and digital compandors", *Bell System Technical Journal*, **49**, No. 7, pp. 1555–1588 (Sept. 1970).
4. *Transmission Systems for Communications*, Fourth Edition, Bell Telephone Laboratories (Feb. 1970).
5. *Waveform Quantization and Coding*, N.S. Jayant (Ed.), New York: IEEE Press, 1976.

Digital Interpolation of Stochastic Signals

A. D. POLYDOROS and E. N. PROTONOTARIOS

Department of Electrical Engineering, National Technical University of Athens, Athens, Greece

1. Introduction

Interpolation between the samples of a digital sequence is a process often required in signal processing, wherever it is necessary to change from one sampling rate to another. The solution of the related problems of interpolating between the data, delaying by a noninteger factor, and changing the sampling rate is useful in the areas of speech processing, modulation systems and coding systems. Schafer and Rabiner[1] have suggested a scheme employing a digital low-pass filter for this sampling rate alteration. In a number of papers published thereafter methods for improved implementation of the low-pass filter were proposed, in terms of minimizing the overall amount of computation.[2,3] In this whole work the key idea is the construction of the fixed low-pass filter which only depends on the desired design characteristics and not on the type of the input digital signal.

In contradistinction to that, the scheme proposed in the present paper employs a digital FIR filter whose impulse response takes into account the known structure of the stochastic input signal. The filter is optimum in the sense that it minimizes the mean-square interpolation error f_{int}, and it requires the knowledge of a certain sampled version of the autocorrelation function $R_{xx}(\tau)$ of the input sequence.

The analysis will follow two steps: first, the interpolation in the middle of the sampling interval is investigated; second, the generalization for an alteration by a noninteger factor follows. Although the former step can be viewed as a special case of the latter, the distinction is necessary because the first problem leads to a particularly simple closed form solution not available in the second case. Finally, the theory is followed by some illustrative examples and the computer simulation results. In this last part, we compare the proposed filter with a classical interpolator (Lagrangian filter) as well as with some other low-pass filters resulting from various design methods.

2. The optimum filter

Figure 1 shows schematically the operation of the interpolating digital filter $H(z)$. Consider the input sequence $x(n)$ which is the T sec-sampled version of the input signal $x(t)$ (the dotted envelope in Fig. 1). The desired output is the sequence $y(n)$ which is actually the T' sec-sampled sequence of $x(t)$. Suppose that $T'=(M/L)T$, where $M, L \in \mathbb{N}$. To achieve this fractional rate we can first increase the sampling rate by a factor of L by interpolating, and then decrease it by a factor of M, by simply extracting one every M successive samples. In the following we are occupied with the interpolating part of the whole process, which is actually the main problem, that is we consider the case $T'=T/L$, $L \in \mathbb{N}$.
Let us define now the sequence $u(n)$ as

$$u(n)=\begin{cases} x(n/L), & n=0,\ \pm L,\ \pm 2L,\ \ldots \\ 0, & \text{otherwise} \end{cases}$$

Actually $u(n)$ is $x(n)$ itself, "enriched" with zero-valued samples at the points where interpolation is desired; $u(n)$ is the "filter input sequence", on which the filter operates to produce $y(n)$. As proved in ref. 1, an ideal low-pass filter with bandwidth $1/2T$ will precisely perform the interpolation process. Of course, only an approximation of the ideal filter is possible by any digital design technique, and the efforts of the authors were concentrated on the problem of the optimal design of the filter, given the tolerances and the ripples of the passband and stopband.[4,5] Although the use of IIR filters almost always results in a reduced amount of necessary computation compared to FIR filters, the use of FIR filters is urged from phase considerations for this specific problem. As a matter of fact, by using a symmetric impulse responce, $h(n)=h(-n)$, we obtain a linear phase FIR filter, which corresponds to a delay of an integer number as samples. In contrast, no IIR filter design technique yields precisely linear phase. Thus with FIR filters, the error due to phase nonlinearity can be zero.

Another comment regards the choice of N, the length of the impulse response. Since an even N corresponds to a delay of at least one half-sample,[5] we avoid this undesirable interpolation by choosing an odd N. To conclude the impulse response constraints, we mention that a reasonable

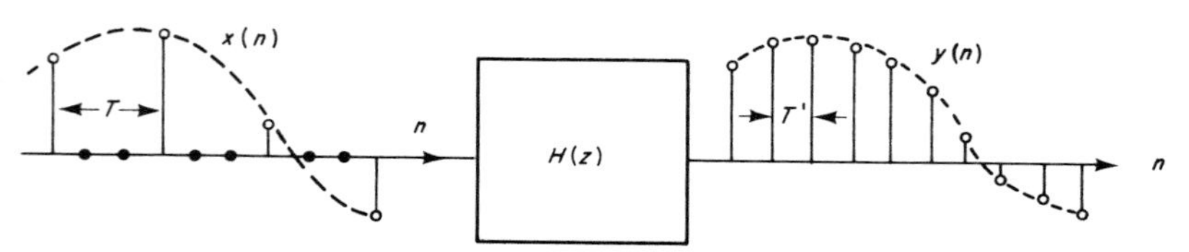

Fig. 1. Interpolating filter.

requirement on the interpolation filter is that the values of the output sequence $y(n)$ at the original sampling times be the same as the original samples. This implies[1] that $h(0)=1$ and

$$h(n)=0, \quad n=\pm \mathrm{L}, \pm 2\mathrm{L}, \ldots, \left[\frac{N-1}{2L}\right]L.$$

What we propose is a filter matched to the statistical structure of the input signal. It specifically requires only the knowledge of $R_{xx}(\tau)$ (or, equivalently, the spectrum $S_{xx}(f)$) for the determination of the coefficients $h(n)$.

We now consider *interpolation at the middle of the sampling interval* ($L=2$). As mentioned before, the criterion of optimization is the minimization of the mean-square interpolation error:

$$f_{\mathrm{int}} \triangleq E\{e^2(nT')\} = E\{|x(nT')-y(nT')|^2\} = \min \tag{1}$$

After some tedious computational details we derive the following expression for f_{int}.

$$f_{\mathrm{int}} = R_{xx}(0) + \sum_{l=-k}^{k} \sum_{m=-k}^{k} R_{xx}\left(\frac{l-m}{2}T\right) h(l)h(m) - 2\sum_{m=-k}^{k} R_{xx}\left(\frac{mT}{2}\right)h(m) \tag{2}$$

where $l, m \in \mathbb{Z}_{\mathrm{odd}}$ ($\mathbb{Z}$: integers) and where the index k is related to the length N by the relation

$$N=2k+1, \quad k \in \mathbb{N}_{\mathrm{odd}} \tag{3}$$

After having determined the error function f_{int}, it is a cumbersome but straightforward task to determine the coefficients which jointly minimize equation (2) and obey the previously mentioned restrictions $h(n)=0$ for $n=$even and $h(-n)=h(n)$. The appropriate method to be followed is to differentiate equation (2) with respect to each of the coefficients and set the result equal to zero. It can be proved that the resulting values minimize f_{int}

Differentiating we find the coefficients $h(n)$ by solving the system of linear equations:

$$A x \boldsymbol{h} = \boldsymbol{B} \rightarrow h = \boldsymbol{A}^{-1} x\, \boldsymbol{B} \tag{4}$$

with typical elements

$$A=\{a_{ij}\}, \quad a_{ij}=R_{xx}[(i-j)T]+R_{xx}[(i+j-l)T]$$

$$B=\{b_i\}, \quad b_i=R_{xx}[(i-\tfrac{1}{2})\mathrm{T}]; \quad \boldsymbol{h}=\{h_i\} \qquad i, j \in (0, 1, 2, \ldots, (k+1)/2)$$

Example:
If $N=3$, then $h(0)=1$ and $h(1)=h(-1)$ is the unique unknown variable in

the expression for f_{int}. In this case

$$h(1)=h(-1)=\frac{R_{xx}(T/2)}{R_{xx}(0)+R_{xx}(T)}$$

This is actually an improved version of the linear interpolation between the samples, for which $h(1)=h(-1)=\frac{1}{2}$, adapted to the statistical structure of the input signal.

We consider next *interpolation at an arbitrary point of the sampling interval* ($L \geqq 2$). Here we generalize the previous results and try to determine the optimum filter which alters the sampling rate by an integer factor greater than two. The filter operates on the "filter input sequence" $u(n)$ and yields at the output a minimum mean-square error approximation of the T'-sec sequence $y(n)$. The criterion of optimization now is slightly different from that of the previous paragraph. What is to be minimized now is the average (arithmetic mean value) of the individual expected square-errors of the interpolation at the $L-1$ points in the interval T, i.e. the minimization of

$$f_{\text{int}}=\frac{1}{L-1}\sum_{\xi=1}^{L-1} f_{\xi,int}$$

where $f_{\xi,\text{int}}$ corresponds to the mean square error at the point ξ. Of course, if only one point in the T-interval were of interest, then the error function to be minimized would include only that point, but such a case does not seem to be of great practical interest, except for the problem of a simple noninteger delay.

In addition to the previous constraints we insert now a new one, namely that $Q=2p$ (even number) samples will always be involved in the computation, and the interpolation will take place at the middle interval, so that there will be $Q/2$ samples at each side of the interpolation interval. This condition assures the symmetry of the impulse response and the linearity of the phase. The length N can be seen to be

$$N=QL-1=2pL-1 \qquad p\in \mathbb{N} \tag{5}$$

After a long computation we find that the error function f_{int} is given by

$$f_{\text{int}}=\frac{1}{L-1}\sum_{\xi=1}^{L-1}\left\{R_{xx}(0)+\sum_{m=pL+\xi}^{(p-1)L+\xi}\ \sum_{l=-pL+\xi}^{(p-1)L+\xi} R_{xx}\left(\frac{l-m}{L}T\right)h(m)h(l)\right.$$
$$\left.-2\sum_{m=-pL+\xi}^{(p-1)L+\xi} R_{xx}\left(\frac{mT}{L}\right)h(n)\right\} \tag{6}$$

where $m, l\in \mathbb{Z}$ and increase by Lunits. After some details we can prove that the coefficients minimizing equation (6) are given by the linear system of equations:

$$\frac{\partial f_{\text{int}}}{\partial h_i}=\frac{4}{L-1}\left\{\sum_{m=-pL+\xi_i}^{(p-1)L+\xi_i} R_{xx}\left(\frac{i-m}{L}T\right)h(|m|)-R_{xx}\left(\frac{iT}{L}\right)\right\}=0 \tag{7}$$

where $m\in \mathbb{Z}$ and increases by L units and ξ_i is the index of the interpolation point in the interval T, $\xi_i\in\{1, 2, \ldots, L-1\}$.

3. Some theoretical examples

We refer now to two simple illustrative examples of the interpolation in the middle of the interval. The chosen stochastic signals have structures which are often encountered in practice at least approximately. For these two categories we can easily determine the optimum filter coefficients for various lengths N and sampling intervals. As a first application, we consider a signal $x(t)$ with an exponential $R_{xx}(\tau)$.

PROPOSITION 1. For a stochastic signal $x(t)$ with exponential autocorrelation function $R_{xx}(\tau)=P_0\exp\{-\omega_0|\tau|\}$, the optimum interpolation filter in the middle of the sampling interval has the coefficients

$$h(0)=1, \quad h(n)=0 \qquad \text{for} \quad n\neq 0, \ \pm 1$$

and

$$h(1)=h(-1)=\exp(-\omega_0 T/2)/\{1+\exp(-\omega_0 T)\}$$

that is, it always has length $N=3$, independently of the original choice of N, and the corresponding mean square interpolation error is $f_{\text{int}}=P_0 \tanh\{\omega_0 T/2\}$ We next assume that the input signal $x(t)$ has a triangular autocorrelation function

$$R_{xx}(\tau)=\begin{cases} P_0\left\{1-\dfrac{|\tau|}{T_0}\right\} & \text{for} \quad |\tau|\leqq T_0 \\ 0 & \text{elsewhere} \end{cases} \tag{8}$$

PROPOSITION 2. For stochastic signals with correlation of the form (8) and sampling interval satisfying the relation $T\leqq 4T_0/(N+1)$ (N the length of the filter) the optimum filter for interpolation in the middle of the interval is exactly the linear interpolator, independently of the original choice of T and N.

4. Computer simulation and results

In this last section we report the computer simulation results and confirm the analytical results obtained above for the middle-interval interpolation algorithm. First we checked that the algorithm functions satisfactorily by evaluating the coefficients of the filter, when the input signal had a "well behaved" $R_{xx}(\tau)$ such as exponential and triangular, for which the theoretical values of $h(n)$ have been calculated as in Section 3. In the simulation part, the signal was generated by the difference equation:

$$x(n+2) - 1{\cdot}1x(n+1) + 0{\cdot}3x(n) = C\xi(n+2); \qquad x(0) = x(1) = 0 \tag{9}$$

where $\xi(n)$ is white Gaussian noise produced by a subroutine, and the constant C regulates the variance of the noise. The corresponding linear system $G(z)$ through which $\xi(n)$ produces $x(n)$ has impulse response $g(n) = 10C\{(0{\cdot}6)^{n+1} - (0{\cdot}5)^{n+1}\}$. $R_{xx}(\tau)$ can be evaluated for this scheme. With $R_{xx}(\tau)$ available, the algorithm determines the coefficients of the optimum filter for $N=11$. These are shown in the first column of Table 1. The other columns contain the values of the coefficients corresponding to filters of the same length $N=11$ as the Lagrange interpolator[1] and the low-pass filter design by various windowing methods, such as the rectangular window, the Kaiser window, the Hamming and Hanning window. The design features are very clearly detailed in refs. 4 and 5. The penultimate row contains the theoretical mean-square error of the interpolation as calculated by the algorithm and the last row contains the results for the mean-square error obtained by the simulation, and from the comparison it is clear that the optimum filter performs better that the others.

5. Conclusion

In this paper we have developed a new approach to the problem of interpolation in a digital sequence and the related problem of changing the sampling rate. Starting from well established work on the subject by previous authors, we shifted the point of view by considering stochastic input signals with given spectra. Formulae or algorithms for the coefficients

TABLE 1. Coefficients for the optimum filter with $N=11$. The signal is given by equation (9) with $C=5$.

Filter	Rectangular window	Kaiser window	Hamming window	Hanning window	Lagrange window	Optimum filter
$h(1)$	0·63661977	0·60138186	0·59013165	0·58608920	0·58593750	0·55852825
$h(3)$	−0·21220659	−0·12246130	−0·10069949	−0·09100322	−0·09765625	−0·07391147
$h(5)$	−0·12732395	0·01901850	0·01255937	0·00257876	0·01171875	0·00893010
Theor.M.S.E.	13·424442	11·567216	11·423011	11·402195	11·382521	11·259595
M.S.E.	12·576160	11·437487	10·974069	10·974900	11·194594	10·329415

of the optimum filter have been given, based on the autocorrelation function of the input signal; computer simulation results verified the validity of the proposed algorithm. In the case that $R_{xx}(\tau)$ is absolutely known, it seems that the algorithm is markedly superior in the class of the known interpolators. A problem which arises concerning the superiority of the interpolator proposed concerns the case where $R_{xx}(\tau)$ is not known and has to be estimated or even guessed. It is then possible that the optimum algorithm might perform worse than some other interpolators, the degredation being the result of the lack of complete information about the input. It seems that the scheme proposed fits much better when there exist constraints on the complexity of the filter, which require small values of N.

References

1. R. Schafer and L. Rabiner, "A Digital Signal Processing Approach to Interpolation," *Proc. IEEE*, **61**, No. 6, pp. 692–702, (June 1973).
2. R. Grochiere, L. Rabiner and R. Shively, "A Novel Implementation of Digital Phase Shifters, *B.S.T.J.*, **54**, No. 8, pp. 1497–1502, (Oct. 1975).
3. R. Grochiere and L. Rabiner, "Optimum FIR Digital Filter Implementations for Decimation—Interpolation and Mark Band Filtering," *Proc. IEEE*, **61**, (Oct. 1975).
4. L. Rabiner and B. Gold, "Theory and Application of Digital Signal Processing," Englewood Cliffs, N.J.: Prentice-Hall, 1975.
5. A. Oppenheim and R. Schafer, "Digital Signal Processing," Englewood Cliffs, N.J.: Prentice-Hall, 1975.

Branch Filtering using FIR and IIR Complementary Structures

T. A. RAMSTAD

Department of Electrical Engineering, Norwegian Institute of Technology, Trondheim, Norway

1. Introduction

In many filtering applications one wishes to separate two signals in adjacent frequency bands. Depending on the filter specifications, different branch filter structures may be efficient. In this paper we shall discuss two approaches for constructing branch filters with complementary filtering properties, the first one based on FIR linear phase filters, the second on IIR filters. One possible application of these structures is in the socalled transmultiplexer[1-4] (The transmultiplexer is a digital equipment for converting the 60 channels in an FDM super-group to 2×30 TDM channels (two 1st order PCM systems)). In the transmultiplexer the task of the branch filter is to separate the speech signal and an out of band signalling frequency at 3825 Hz in a sampled version ($f_s = 8$ kHz) of the combination of the two signals. There is a variety of other possible applications, but the efficiency of the structures is strongly dependent on filter requirements and bandwidth of the two signals as will be apparent from the discussion. A branch filter structure based on wave digital filters has formerly been proposed in ref. 5.

2. FIR-complementary filter implementation

The FIR-complementary filter structure is based on a very simple transformation which interchanges stopbands and passbands in the linear phase filter as well as the tolerances of the bands.

Consider a filter $H(z)$, as in Fig. 1, which suppresses the signal in a given frequency band. If the output of the filter is subtracted from a delayed version of the input (the delay must be equivalent to the filter delay), the

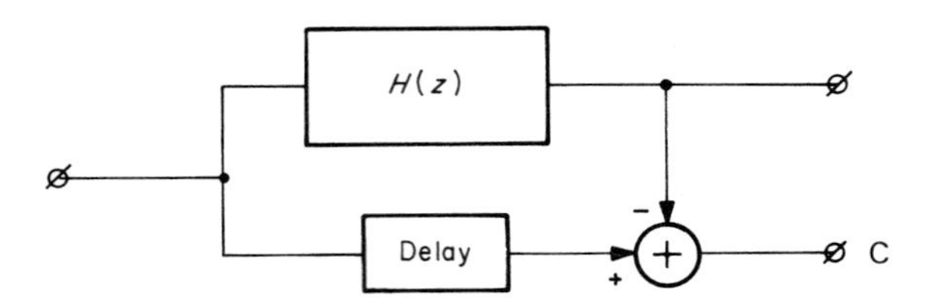

Fig. 1. FIR filter with complementary output (C).

complementary signal is obtained. If the transfer function of the filter is given by

$$H(z)=\sum_{n=0}^{N-1} h(n)Z^{-n} \tag{1}$$

and N is odd, then the transfer function from the input to the complementary output can be expressed as

$$C(Z)=Z^{-(N-1)/2}-\sum_{n=0}^{N-1} h(n)Z^{-n} \tag{2}$$

It is easily seen that the two responses are related through

$$C^*(e^{j\omega})=1-H^*(e^{j\omega})$$

where the star operator indicates that the linear phase factors are omitted. As an example, the responses of two complementary filters are presented in Fig. 2.

As can be observed from Fig. 2, the deviation from the nominal values in the bands has changed from the stopband in one filter to the passband in the other and vice versa. This method can for instance be utilized for LP–HP-transformation by simply subtracting 1 from the centre coefficient of the filter. (When the coefficients are properly normalized). More important is the possibility of extracting the complementary signal by simply introducing one extra addition as shown in Fig. 3. An example of such a filter is presented in Section 4.

3. IIR-complementary filter implementation

Because of the non-linear phase of an IIR filter one cannot generally obtain the complementary signal as easily as was done in the preceding paragraph with FIR linear phase filters. Obviously, one solution to the problem is to introduce a frequency dependent delay in Fig. 1 which compensates for the non-linear phase behaviour of the filter. However, the delay, which may be realized as an allpass filter, has a tendency to become as complex as would

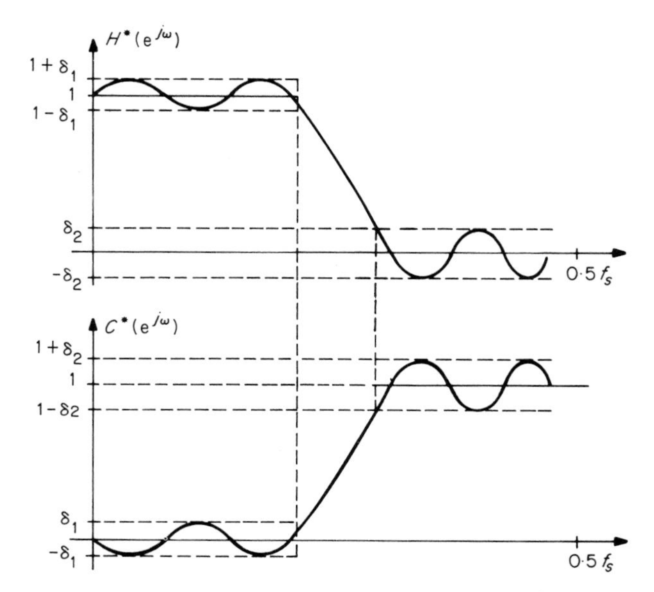

Fig. 2. Connection between the tolerances of $H(Z)$ and $C(Z)$.

be a realization of the same system with two filters. Only in the case where one of the signals is narrow band, does an approach like this seem appropriate.

We shall propose a structure which is based on the transmission and reflection coefficients of a lossless filter. The transmission and reflection

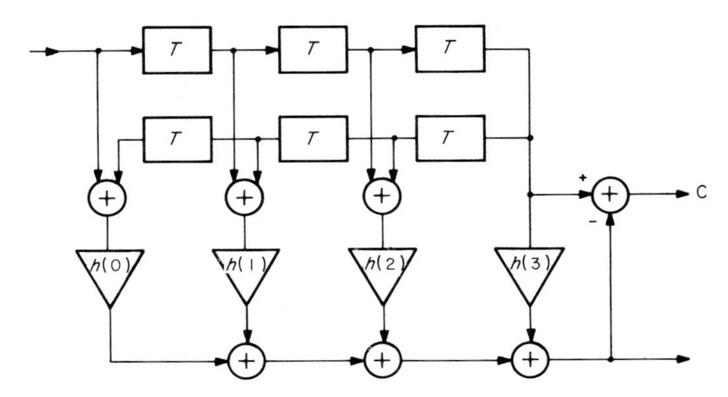

Fig. 3. Structure for implementing the complementary output of a linear phase FIR filter (here $N=7$).

coefficients have the same poles, but different zeros. The two outputs will share the same register for the intermediate results in the filter, but will have different coefficients realizing the zeroes of the two branches. Both direct from II structures and parallel structures can be implemented.

We shall denote the transfer function of the two branches of the filter as $H(Z)$ and $C(Z)$. Furthermore, $H(Z)$ is written as a rational function

$$H(Z)=\frac{T(Z)}{N(Z)} \tag{4}$$

where the polynominal $T(Z)$ has all its zeros on the unit circle due to the constraint that the filter be lossless. (This corresponds to an LC analog filter). The complementary filter can be described by

$$C(Z)=\frac{P(Z)}{N(Z)} \tag{5}$$

where $N(Z)$ is common in equations (4) and (5).

The filters will be complementary if the frequency responses obey the following equation

$$|C(e^{j\omega})|^2=1-|H(e^{j\omega})|^2 \tag{6}$$

This relation leads to the tolerances for $C(Z)$ as given in Fig. 4 when the tolerances in $H(Z)$ are given by Fig. 2(a). The error tolerances in $C(Z)$ are thus

$$\delta_2^C=[1-(1-\delta_1^H)^2]^{1/2} \tag{7}$$

$$\delta_1^C=[1-(\delta_2^H)^2]^{1/2} \tag{8}$$

An analytic continuation of equation (6) leads to a relationship between

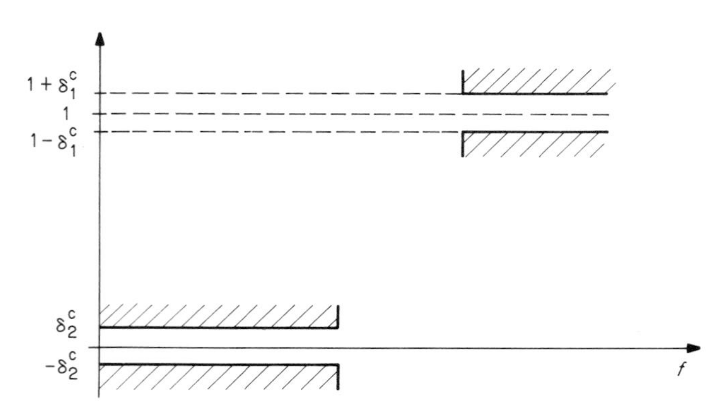

Fig. 4. Error tolerance for the complementary filter when constructed from $H(Z)$ in Fig. 2(a) and equation (6).

the transfer-functions of the two branches of the filter

$$C(Z)C(Z^{-1}) = 1 - H(Z)H(Z^{-1}) \tag{9}$$

which upon the insertion of equations (4) and (5) yields the expression for the zeros of $C(Z)$ as

$$P(Z)P(Z^{-1}) = N(Z)N(Z^{-1}) - T(Z)T(Z^{-1}) \tag{10}$$

We shall restrict the further discussion to filters with Chebycheff tolerances in the passband. In this case $P(Z)$ can be realized with real coefficients and minimum degree. To use equation (10) directly the amplification factor of $H(Z)$ must be suitably normalized (to give maximum amplification equal to 1). This is due to the fact that the complementary filter will require transmission zeros at the frequencies where the frequency response of $H(Z)$ crosses 1. To split $P(Z)P(Z^{-1})$ to find $P(Z)$, it is necessary to have double zeros of $1-|H(e^{i\omega})|^2$. This can be obtained if the situation in Fig. 5 exists. Accordingly, $C(Z)$ will have transmission zeros where $H(Z)$ has maximum values in the passband. This observation is incidentally an alternative way of determining $P(Z)$ in the complementary filter.

$H(Z)$ and $C(Z)$ can be realized as a direct form II structure using the same recursive part for both filters. We shall however present only the parallel realization as an example. In that case the two transfer functions are expressed by

$$H(Z) = \sum_{k=1}^{M/2} \frac{\alpha_k + \beta_k Z^{-1}}{1 + r_k Z^{-1} + p_k Z^{-2}} + h \tag{11}$$

and

$$C(Z) = \sum_{k=1}^{M/2} \frac{\sigma_k + \gamma_k Z^{-1}}{1 + r_k Z^{-1} + p_k Z^{-2}} + c \tag{12}$$

when the number of poles is not less than the number of zeros. (One of the β_k, γ_k, and p_k will be zero if the filter order is odd leading to one first-order section). The structure is shown in Fig. 6. An example of an IIR branch filter is given in the next section.

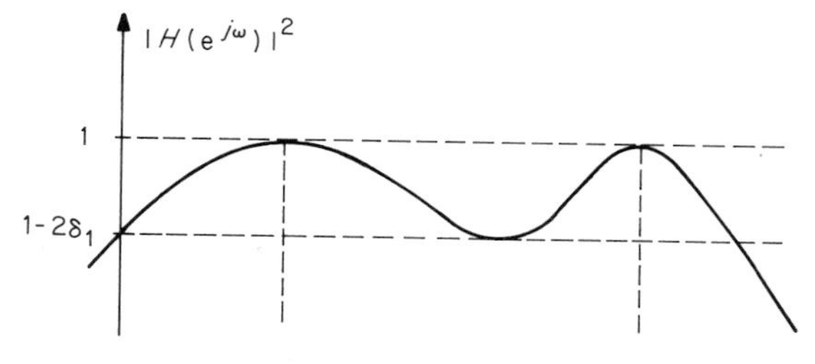

Fig. 5. Passband when the maximum amplification in an elliptic filter is 1.

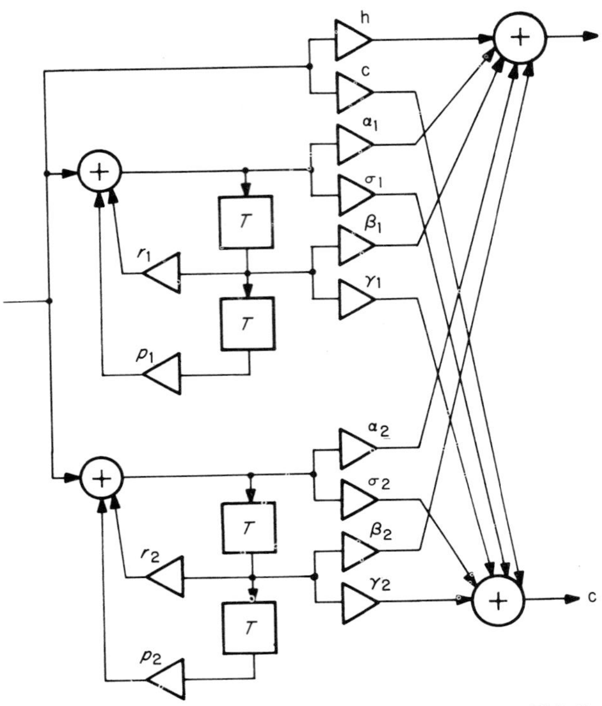

Fig. 6. Implementation of an IIR branch filter with parallel structure. This figure shows a filter of fourth order.

4. Examples of filter implementation

In the following examples we shall design branch filters of both the FIR and IIR type. They will be specified to separate the speech signal and the out-of-band signalling tone at 3825 Hz in a sampled FDM signal. The requirements for the speech channel and the signalling frequency channel are given in Fig. 7.

For the FIR case we combine the two requirements leading to error tolerances as given in Fig. 8. The filter is designed using the linear phase FIR optimization program in ref. 6. The resulting FIR filter of length 37 is presented with its coefficients in Table 1 and the frequency responses of $H(Z)$ and $C(Z)$ are given in Fig. 9. (The coefficients of $C(Z)$ are equivalent except for $h_c(19)$ which is given by $h(19)-1$).

When constructing an IIR branch filter, the tolerance requirements in Fig. 7 are combined keeping in mind relations (7) and (8). The resulting tolerance scheme is shown in Fig. 10.

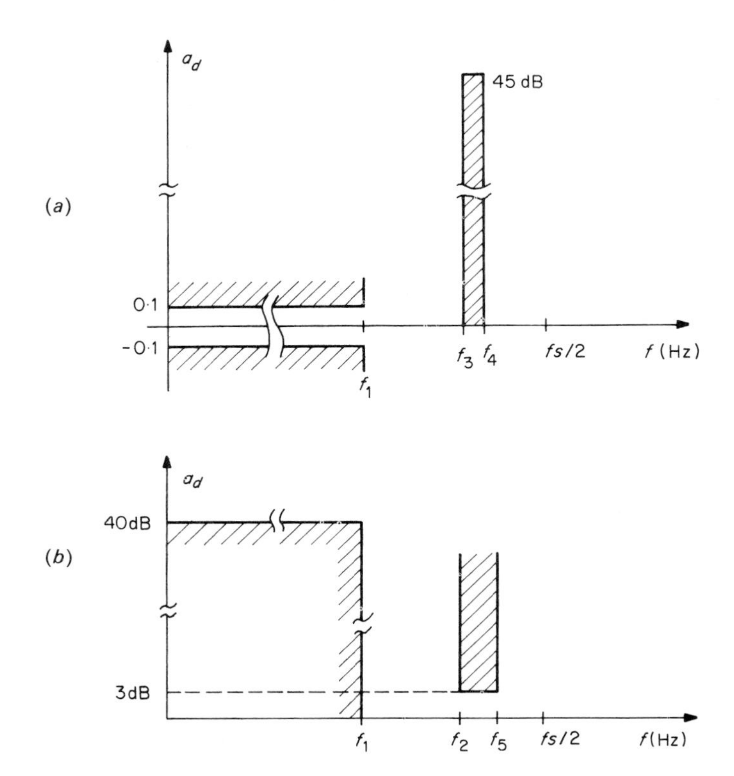

Fig. 7. Filter tolerances. (*a*) Suppression of signalling from speech. (*b*) Suppression of speech from signalling. $f_1 = 3400$ Hz; $f_2 = 3809$ Hz; $f_3 = 3819$ Hz; $f_4 = 3831$ Hz; $f_5 = 3841$ Hz; $f_s = 8000$ Hz.

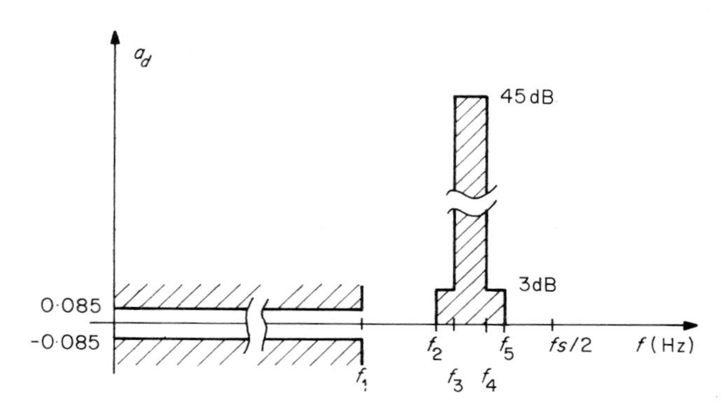

Fig. 8. Error tolerances for the FIR complementary structure.

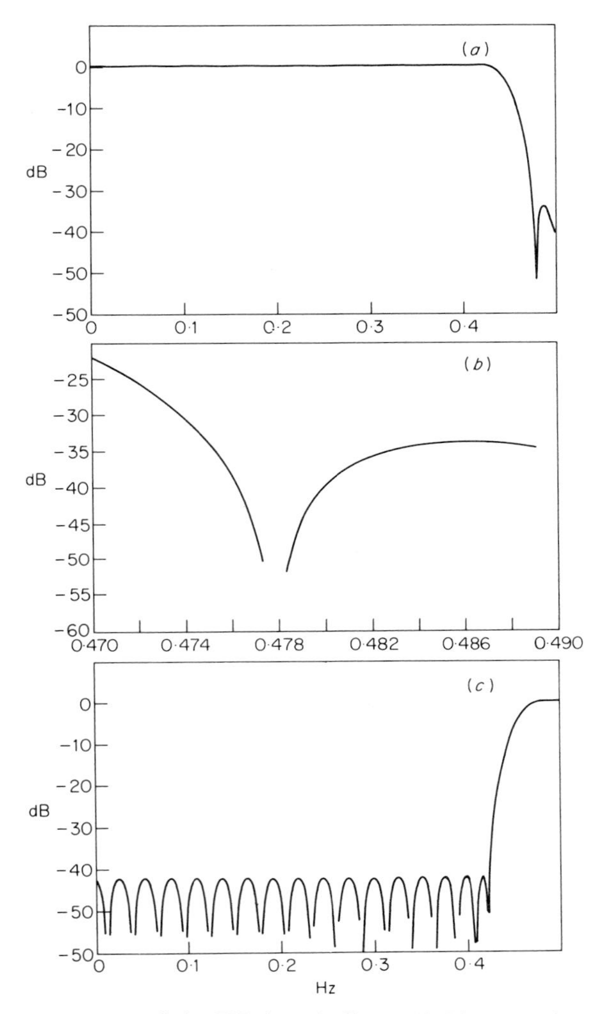

Fig. 9. Frequency responses of the FIR branch filters. (*a*) The gross frequency response of *H*(*Z*). (*b*) Details of the stopband of *H*(*Z*). (*c*) Response of *C*(*Z*). The frequency scale is normalized with respect to the sampling frequency (8 kHz).

For convenience we have not designed an optimum filter but rather an elliptic filter which has its first transmission zero at 3825 Hz satisfying the required 45 dB attenuation in a 12 Hz band. The stopband is extended to 4000 Hz with an equal ripple of approximately 36 dB attenuation. The coefficients of this filter and its complementary counterpart are given below

TABLE 1. Filter coefficients of the FIR speech channel filter.

Impulse Response
$h(1) = 0{\cdot}64010802\text{–}02 = h(37)$
$h(2) = -0{\cdot}58875901\text{–}02 = h(36)$
$h(3) = 0{\cdot}77628682\text{–}02 = h(35)$
$h(4) = -0{\cdot}92403176\text{–}02 = h(34)$
$h(5) = 0{\cdot}99096850\text{–}02 = h(33)$
$h(6) = -0{\cdot}93416050\text{–}02 = h(32)$
$h(7) = 0{\cdot}71358161\text{–}02 = h(31)$
$h(8) = -0{\cdot}29767963\text{–}02 = h(30)$
$h(9) = -0{\cdot}33099154\text{–}02 = h(29)$
$h(10) = 0{\cdot}11720610\text{–}01 = h(28)$
$h(11) = -0{\cdot}22045220\text{–}01 = h(27)$
$h(12) = 0{\cdot}33855322\text{–}01 = h(26)$
$h(13) = -0{\cdot}46538287\text{–}01 = h(25)$
$h(14) = 0{\cdot}59329270\text{–}01 = h(24)$
$h(15) = -0{\cdot}71384772\text{–}01 = h(23)$
$h(16) = 0{\cdot}81854225\text{–}01 = h(22)$
$h(17) = -0{\cdot}89963796\text{–}01 = h(21)$
$h(18) = 0{\cdot}95095881\text{–}01 = h(20)$
$h(19) = 0{\cdot}90314751\text{–}00 = h(19)$

in Table 2. (The designation of the coefficient is according to equations (11) and (12) and Fig. 6).

The frequency responses of $H(Z)$ and $C(Z)$ are given in Fig. 11.

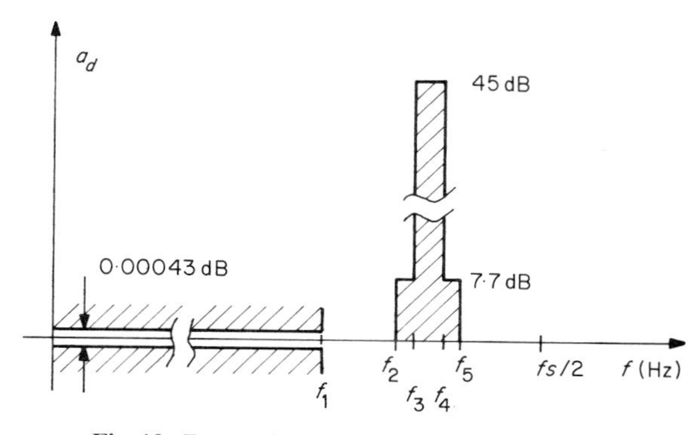

Fig. 10. Error tolerances of the IIR branch filter.

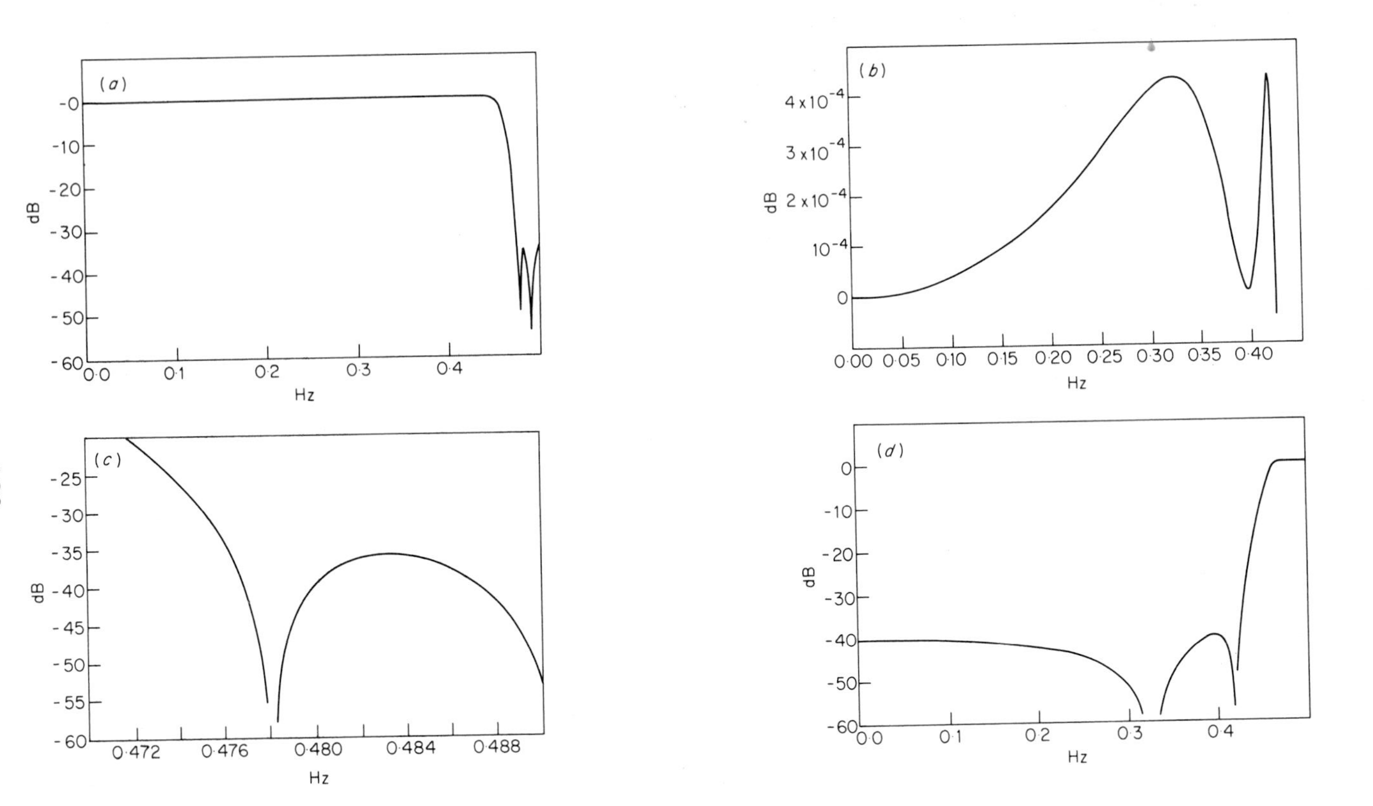

Fig. 11. (*a*) The gross frequency response of $H(Z)$. (*b*) Details of the passband of $H(Z)$. (*c*) Attenuation in the neighbourhood of 3825 Hz for $H(Z)$. (*d*) Frequency response of $C(Z)$. The frequency scale is normalized with respect to the sampling frequency (8kHz).

TABLE 2. Coefficient of $H(Z)$ and $C(Z)$ in the structures in (11) and (12).

$r_1 =$	1·598036	$p_1 =$	0·648192
$r_2 =$	1·807593	$p_2 =$	0·867436
$h =$	1·33894		
$\alpha_1 =$	−0·507138	$\beta_1 =$	−0·418997
$\alpha_2 =$	−0·086068	$\beta_2 =$	−0·111107
$C =$	0·009571		
$\sigma_1 =$	0·138764	$\gamma_1 =$	0·060953
$\sigma_2 =$	−0·147373	$\gamma_2 =$	−0·113927

5. Conclusion

In the preceding paragraphs we have introduced complementary realizations of branch filters of both FIR and IIR types, and examples of both structures have been given.

These structures seem to be efficient when the band tolerances of the complementary realization do not far exceed the band tolerances of the original filters. This is most critical for the IIR case because the stopband requirement in one filter leads to extreme passband requirements in the complementary filter through equation (8).

The method presented may also be applied to multiband filters.

Acknowledgment

This work has been supported by the Norwegian Telecommunication Administration and the Norwegian Defence Communication Administration as a part of the developments of the so called transmultiplexer.[1] I express my thanks to Dr.ing. T. Røste for inspiring discussion on the subject of the paper.

References

1. H. Holt and E. Nanseth., "Transmultiplekser. Et nytt utstyr for konvertering mellom PCM og FDM, Del 1; Anvendelser og grunnprinsipper". *Telektronikk*, No. 4, pp. 362–368 (1977). (In Norwegian).
2. T. Ramstad and E. Nanseth. "The Transmultiplexer. A New Type of Equipment for Signal Conversion between PCM and FDM Systems, Part II: Theory", *Telectronikk*, No. 2, pp. 131–137 (1979).

3. A. Anderson, T. Røste and P. Wilhelmsen, "The Transmultiplexer. A New Type of Equipment for Signal Conversion between PCM and FDM Systems, Part III: Specifications", *Telektronikk*, No. 1 (1968).
4. O. Drageset and O. Foss: "The Transmultiplexer. A New Type of Equipment for Signal Conversion between PCM and FDM Systems, Part IV: Implementation", *Telektronikk*, No. 1 (1978).
5. M. Bellanger *et al.*, "Un Filtre Numérique d'Onde Complémentaire pour Voie Téléphonique," Cables & Transmission, 31ᵉ A., No. 4 (Oct. 1977). (In French).
6. J. H. McClellan *et al.*, "A Computer Program for Designing Optimum FIR Linear Phase Digital Filters," *IEEE Trans. Audio Electroacoustics* **AU-21**, pp. 506–526, (Dec. 1973).

Two-dimensional Kalman Filtering with Applications to the Restoration of Scintigraphic Images

L. BARONCELLI,† A. DEL BIMBO† AND G. ZAPPA

Istituto di Elettronica, Facoltà di Ingegneria University of Florence, Florence, Italy

1. Introduction

The aim of this paper is to provide an introduction to the application of Kalman filtering techniques to image restoration, to analyse some numerical problems connected with their implementation and to develop an efficient algorithm which will be applied to the restoration of scintigraphic images, an important diagnostic tool of nuclear medicine.

In the last few years these techniques have received considerable attention for several reasons. In the first place for the increasing number of fields (like remote sensing, nuclear medicine, astronomy, etc.) where digital image processing was successfully applied. Then for the great computational advantages of recursive algorithms in the spatial-domain over usual frequency-domain filtering techniques. Finally, since the statistical approach to image processing constitutes a useful basis on which to approach related problems like adaptive filtering, boundaries detection and pattern recognition. Moreover the researchers' interest was attracted by the theoretical problems which arise when results on modelling, identification and recursive estimation of one-dimensional (1-D) processes are extended to multi-dimensional processes.

Despite the numerous contributions which have appeared on this subject, in our opinion, a lot of work has still to be done and a comprehensive theory of two-dimensional recursive filtering is still lacking. Hence we derive a 2-D Kalman filter only for a particular class of models and in the most direct way, viz. exploiting 1-D results. Our approach is essentially based on the development presented in ref. 4. Only some modifications concerning the admissible models and numerical algorithms are introduced. A new strip-processing technique is also suggested.

† Present address: IBM Italia, Milano, Italy.

This paper is organized as follows. The restoration problem and the Wiener filter solution are briefly presented in Section 2 together with some inherent limitations. In Section 3 a 2-D Kalman filter is developed and in Section 4 its application to the restoration of scintigraphic images is studied and the results are discussed. Conclusions are presented in Section 5.

2. Image restoration

The fundamental model describing the image restoration problem is given by

$$z(i, j)=\sum_{k,l} h(i-k, j-l)y(k, l)+v(i, j) \tag{1}$$

where $z(.,.)$ and $y(.,.)$ are the recorded and the original image respectively, both sampled on an equally spaced array of points indexed by (i, j), $v(.,.)$ is an additive corrupting noise and $h(.,.)$ is the sampled point-spread-function (PSF) representing the degradation introduced by the image acquisition system. This degradation is supposed linear and spatially invariant.

In terms of equation (1) the image restoration problem can be stated as that of estimating the original image y from the observed image z, assuming that the PSF h is known and exploiting any statistical information available about y and v.

Under the usual assumption that y and v are independent, space-invariant, zero-mean Gaussian processes with known covariances r_y and r_v, then the image restoration problem can be solved through a minimum-mean-square-error (MMSE) estimate of y provided by a Wiener filter (WF). This filter is conveniently described in the 2-D z-transform domain. Indeed, denoting by $H(u, v)$, $R_y(u, v)$ and $R_v(u, v)$ the z-transforms of $h(i, j)$, $r_y(i, j)$ and $r_v(i, j)$ respectively, we have

$$\hat{Y}(u, v)=\frac{H(u, v)^*}{|H(u, v)|^2+[R_v(u, v)]/R_y(u, v)]}Z(u, v) \tag{2}$$

where $Z(u, v)$ and $\hat{Y}(u, v)$ are the z-transforms of the observed and the estimated images.

The WF however has several limitations as a image processing system.[1] In fact, MMSE estimates are mainly concerned with noise suppression to the detriment of image resolution. Moreover, in order to improve computational efficiency, we are forced to assume stationarity, which provides a further reduction of contrast and smoothing of edges. Furthermore for large images the computational burden is high, in terms of both memory requirements and processing time, since the entire image is processed at the same time. Even FFT algorithms and various partitioning techniques do not make the WF suited for a fast implementation on a minicomputer.

While other convolutional filters have been designed for increasing image resolution (inverse filter, constrained least-square filter etc.[2]) great computational advantages can be obtained by exploiting recursive techniques in the space-domain like the Kalman filter (KF) which has had so many successful applications in 1-D recursive estimation problems. Moreover within a recursive scheme non-space-invariant processes can be more easily handled.

However the extension of KF algorithms to 2-D processes is far from being trivial and many theoretical and numerical problems must still be solved. We list below some of the difficulties to be overcome.

(1) On a 2-D array there is not a natural order among the points, and, consequently, various directions of recursion can be defined, each requiring a different set of initial conditions. Thus we can have, for instance, one-quadrant recursive filters or half-plane recursive filters.
(2) As in the 1-D case, a state-space model for the image is necessary. But in the 2-D case many state-space models can be introduced and, up to now, no general realization theory exists even for deterministic systems.
(3) As a consequence of the absence of the so-called fundamental theorem of algebra for two-variable polynomials, no spectral factorization theorem exists for rational power-density functions of 2-D processes.[3] Therefore, even for a low-dimensional linear stochastic dynamical system, no finite-dimensional innovation representation could exist.
(4) It is difficult within a state-space model of low dimension to represent the effects of the PSF.
(5) So far, a satisfying identification theory has not yet been established for 2-D processes. Thus the parameters of any state-space model of the image are estimated in a rather empirical way and the filtered image could be quite sensitive to these parameters.

These points indicate how many different recursive algorithms have been introduced, each based on different assumptions, and that a general framework is still lacking.

3. A 2-D Kalman filter

According to the previous discussion, we shall limit ourselves to consider a particular class of models, that look suitable for a straightforward implementation of a half-plane recursive KF. Moreover we shall assume that no blur affects the image, viz. the PSF reduces to a Kronecker delta function. More precisely we make the following assumptions

Assumption 1. The observed image $z(i, j)$, $(i, j) \in N \times M$, can be decomposed as

$$z(i, j) = y(i, j) + v(i, j)$$

where y, representing the image to be recovered, is a zero-mean, space-invariant, Gaussian process with covariance

$$r_y(k, l) = E[y(i+k, j+l)y(i, y)]$$

and v is a zero-mean, space-invariant, Gaussian white-noise process with variance θ, uncorrelated with y.

Assumption 2. The process y can be represented by the following state-space model

$$y(i, j) = Hx(i, j) \qquad x(i, j+1) = Fx(i, j) + w(i, j) \tag{3}$$

where $\dim x = \dim w = n$, $\dim H = 1 \times n$, $\dim F = n \times n$, and F is a stable matrix; that is, all its eigenvalues are in modulus less than 1. Moreover x and w are stochastic processes such that

$$\begin{aligned} E[w(i+k, j+l)w(i, j)'] &= \delta_{0,l}Q(k) && \text{with } Q(-k) = Q(k)' \\ E[x(i+k, j+l)w(i, j)'] &= 0 && \text{if } l \geqq 0 \\ E[x(i+k, j)x(i, j)'] &= P(k) && \text{with } P(-k) = P(k)'. \end{aligned} \tag{4}$$

From equations (4) it follows that $P(k)$ and $Q(k)$ satisfy the Liapunov equation

$$P(k) = FP(k)F' + Q(k)$$

Note that equations (3) imply that any line of y has a 1-D horizontally invariant state space model and equations (4) that the state vectors and the noises of these models are correlated in a vertically invariant way.

The model (3) includes the more usual state-space model[4]

$$y(i, y) = Hx(i, j)$$
$$x(i, j) = Fx(i, j-1) + F_1x(i-1, j) - FF_1x(i-1, j-1) + u(i-1, j-1)$$

with $FF_1 = F_1F$, and where u is a white noise process with variance Q_1 and x is a space-invariant process with variance P. In fact, introducing the process

$$w(i, j) = x(i, j) - F_1x(i, j-1)$$

we get model (3) with

$$P(k) = F_1^kP \quad \text{and} \quad Q(k) = F_1^kQ \quad \text{for } k \geqq 0$$

where Q is the solution of the Liapunov equation

$$Q = F\ Q\ F' + Q_1$$

From equation (3) and (4), the covariance function of y satisfies

$$\begin{aligned} r_y(k, l) &= HF^lP(k)H' && \text{for } l \geqq 0,\ k \geqq 0 \\ r_y(k, l) &= HF^lP(k)'H' && \text{for } l \geqq 0,\ k \leqq 0 \end{aligned}$$

Conditions on r_y that guarantee the existence of (3) are summarized in the following theorem, whose proof is reported in ref. 5.

THEOREM. A space-invariant stochastic process $y(i, j)$ has a state-space representation (3), (4) if and only if its covariance function $r_y(k, l)$ can be factored as $r_1(l) \cdot r_2(k)$ and the sequence $r_1(l) r_2(0)$, has a 1-D stable stochastic realization.

Recently more general models, representing also processes with nonseparable covariance function or non trivial PSF have been considered.[6,7]

Now, defining the "large" one-index vectors

$$Z(j) = [z(1, j), \ldots, z(N, j)]' \qquad j = 1, 2, \ldots, M$$

and $V(j)$, $Y(j)$ $X(j)$ and $W(j)$ in an analogous manner, we can rewrite equations (3) and (4) as a 1-D state-space model†

$$Z(j) = \boldsymbol{H} X(j) + V(j) \qquad X(j+1) = \boldsymbol{F} X(j) + W(j) \tag{5}$$

where

$$\boldsymbol{H} = \begin{bmatrix} H & & & \bigcirc \\ & H & & \\ & & \ddots & \\ \bigcirc & & & H \end{bmatrix} \qquad \dim \boldsymbol{H} = N \times Nn$$

$$\boldsymbol{F} = \begin{bmatrix} F & & & \bigcirc \\ & F & & \\ & & \ddots & \\ \bigcirc & & & F \end{bmatrix} \qquad \dim \boldsymbol{F} = Nn \times Nn$$

Moreover

$$E[X(j)\ X(j)'] = \boldsymbol{P}, \qquad E[W(j)\ W(j)'] = \boldsymbol{Q}$$

where

$$\boldsymbol{P} = \begin{bmatrix} P(0) & P(1)' & P(2)' & \cdots \\ P(1) & P(0) & & \\ P(2) & & \ddots & \\ \vdots & & & \end{bmatrix} \qquad \boldsymbol{Q} = \begin{bmatrix} Q(0) & Q(1)' & Q(2)' & \cdots \\ Q(1) & Q(0)' & & \\ Q(2) & & \ddots & \\ \vdots & & & \end{bmatrix}$$

Hence $\boldsymbol{P}$ and $\boldsymbol{Q}$ are symmetric block Toeplitz matrices whose generating function is respectively $P(k)$ nad $Q(k)$.[8]

---

† In the following we shall denote by *bold italic letters* (e.g. $\boldsymbol{H}$) the matrices referring to the state model (5) i.e. those matrices whose dimensions are multiple of N.

Now the 1-D state-space model (5) admits the Kalman filter

$$\hat{X}(j)=\boldsymbol{F}\hat{X}(j-1)+\boldsymbol{K}(j)[Z(j)-\boldsymbol{HF}\hat{X}(j-1)] \qquad \hat{X}(0)=0 \tag{6}$$

where the $nN \times N$ Kalman gain matrix $\boldsymbol{K}(j)$ is obtained by solving the following $(nN \times nN)$ matrix Riccati equation

$$\boldsymbol{K}(j)=\Pi(j)\boldsymbol{H}'[\boldsymbol{H}\Pi(j)\boldsymbol{H}'+\theta I]^{-1} \tag{7a}$$

$$\Pi(j+1)=\boldsymbol{F}\Pi(j)\boldsymbol{F}'-\boldsymbol{F}\Pi(j)\boldsymbol{H}'[\boldsymbol{H}\Pi(j)\boldsymbol{H}'+\theta I]^{-1}\boldsymbol{H}\Pi(j)\,\boldsymbol{F}'+\boldsymbol{Q} \qquad \Pi(0)=\boldsymbol{P} \tag{7b}$$

In this way $\hat{Y}(j)=\boldsymbol{H}\hat{X}(j)$ is the MMSE estimate of the jth column of the original image based upon the first j columns on the left of the observed image. Obviously an optimal smoother, i.e. an estimate of $Y(j)$ based upon the whole observed image can be obtained by combining two KFs, the first evolving from left to right, the second from right to left.

Notice that the diagonal structure of the matrix $\boldsymbol{F}$ in equations (5) decouples our 2-D estimation problem into two 1-D problems. First compute the one-column-ahead prediction of $x(i, j)$

$$E[x(i, j)|Z(j-1), Z(j-2), \ldots]=F\hat{x}(i, j-1)$$

Then update this estimate on the basis of the observation of the jth column by evaluating the ith component of the product between the Kalman gain $\boldsymbol{K}(j)$ and the "innovation" vector:

$$\varepsilon(j)=Z(j)-\boldsymbol{HF}\hat{X}(j-1).$$

Nevertheless, equations (6) and (7) cannot be directly implemented since they involve multiplications and inversions of large matrices. Notice that, in most applications, the column length N varies from 64 up to 512 or more.

A remarkable saving in computations, according to Attasi[4], can be obtained by exploiting the structure of all the matrices involved in (6) and (7). In fact, since they are block Toeplitz matrices, we can exploit the isometric isomorphism between the algebra of bounded Toeplitz operators and the algebra of complex bounded functions on the unit circle of the complex domain. Hence we consider the isomorphic image of (7a) and (7b) in the z-domain:[4]

$$\boldsymbol{k}_j(z)\pi_j(z)H'[H\pi_j(z)H'+\theta]^{-1} \tag{8a}$$

$$\pi_{j+1}(z)=F\pi_j(z)F'-F\pi_j(z)H'[H\pi_j(z)H'+\theta]^{-1}H\pi_j(z)F'+\boldsymbol{q}(z) \tag{8b}$$

where $\boldsymbol{k}_j(z)$, $\pi_j(z)$ and $\boldsymbol{q}(z)$ are respectively the (two-sided) z-transform of the defining sequences of $\boldsymbol{K}(j)$, $\Pi(j)$ and $\boldsymbol{Q}$. Thus at every step j the required gain $\boldsymbol{K}(j)$ is obtained by antitrasforming the function $\boldsymbol{k}_j(z)$, evaluated recursively through equations (8) on a certain number of points of the unit circle of the complex domain.

Obviously a further computing reduction is achieved if we use only the asymptotic Kalman gain $\boldsymbol{K}=\lim_{j\to\infty}\boldsymbol{K}(j)$. In this case it is necessary to solve just an algebraic Riccati equation in $\pi(z)$ instead of several recursive equations (8*b*). This will provide some degradation of the estimate of the first columns of the image.

Since the Kalman gains can be precomputed, processing time depends on the implementation of equation (6), hence, $\boldsymbol{F}$ and $\boldsymbol{H}$ being diagonal, on the evaluation of the product between the matrix $\boldsymbol{K}$ and the innovation vector ε†. Since $\boldsymbol{K}$ is a Toeplitz matrix, this product is a convolution. More precisely, if $\{\ldots, K(-1), K(0), K(1), \ldots\}$ is the defining sequence of $\boldsymbol{K}$, denoting by $(\varepsilon)_i$ the ith component of ε we have

$$(\boldsymbol{K}.\varepsilon)_i=\sum_{\sigma=1}^{i-1} K(\sigma)(\varepsilon)_{i-\sigma}+K(0)(\varepsilon)_i+\sum_{\sigma=1}^{N-i} K(-\sigma)(\varepsilon)_{i+\sigma}$$

Now exploiting 1-D system realization theory,[9] we get the following theorem.

THEOREM. Assume that there exist two quadruples of matrices $\{A^+, B^+, C^+, D\}$ and $\{A^-, B^-, C^-, D\}$, of appropriate dimensions, such that

$$K(0)=2D, \qquad K(i)=C^+(A^+)^{i-1}B^+, \qquad K(-i)=C^-(A^-)^{i-1}B^- \quad i>0 \tag{9}$$

then the product $\boldsymbol{K}.\varepsilon$ can be computed through

$$(\boldsymbol{K}.\varepsilon)_i=\xi_i^+ + \xi_i^- \tag{10}$$

where

$$\xi_i^+ = C^+\eta_i^+ + D(\varepsilon)_i \qquad \eta_{i+1}^+ = A^+\eta_i^+ + B^+(\varepsilon)_i \qquad \eta_1^+ = 0 \tag{11}$$

$$\xi_i^- = C^-\eta_i^- + D(\varepsilon)_i \qquad \eta_i^- = A^-\eta_{i+1}^- + B^-(\varepsilon)_{i+1} \qquad \eta_N^- = 0 \tag{12}$$

In general equation (9) does not hold exactly, unless we adopt very large matrices. Thus, to get some computational benefits, we must adopt an approximate realization of equation (9), i.e. quadruples that match only the first m values, $m \ll N$, of the impulse response $\boldsymbol{K}(\sigma)$. In this way the contributions to the estimate of the ith row from the rows $i-m, i-m+1, \ldots, i+m-1, i+m$ are exactly computed while contributions from other rows are approximated. On the other side, for each point of the image the number of necessary products reduces from $O(N)$ to $O(m)$, if canonical realizations for the quadruple $\{A, B, C, D\}$ are adopted.[5]

We can now summarize the various steps of this algorithm.

† In order to simplify notation, we shall omit the index j.

Off-line

(1) Construct a state-space model for the image on the basis of its autocovariance function.

(2) Evaluate the Kalman gain $\boldsymbol{K}$ and an approximate realization of its defining sequence (cf. equation (9)).

On-line

(3) Given $\hat{X}(j-1)$, compute the one-column-ahead prediction of $X(j)$, $\boldsymbol{F}\,\hat{X}(j-1)$.

(4) Evaluate the innovation vector $\varepsilon(j)=Z(j)-\boldsymbol{H}\,\boldsymbol{F}\hat{X}(j-1)$

(5) Update the estimate of $X(j)$ computing the product $\boldsymbol{K}\cdot\varepsilon(j)$ exploiting the realized subsystems of equations (10)–(12).

Notice that, at expense of a slight approximation in step (2), our 2-D recursive estimation problem has been reduced to two 1-D recursive estimation problem (steps (3) and (5)). The approximation introduced in step (2) confirms that, in general, a finite-dimensional innovation representation for the stochastic process $z(i, j)$, modelled by equations (3), does not exist.

4. Applications

The filtering algorithm developed in the previous section was applied to the restoration of scintigraphic images in order to test its numerical efficiency and its performance as a restoration technique of very noisy images.

Scintigraphy is a technique used in nuclear medicine to map internal organs of the human body. The gamma rays emitted by a radioactive tracer introduced into the organs are detected by a large scintillation crystal. The light flashes are then detected by an array of phototubes, whose signal is amplified until a digital image sampled on a matrix of points, can be represented by some display unit. The quality of scintigraphic images is, in general, very poor due to disturbances of various kinds. Some of them, such as the nonideal response of the focusing system or the phototube adjustment errors can be modelled by a PSF. Others, such as statistical fluctuations of radionuclides, photomultipliers noise, environment radioactivity, etc., have a purely random character.

Since the noise intensity is very high, it seems reasonable to be concerned only with noise reduction, ignoring any blur introduced by the PSF, at least in a preliminary stage. Moreover some of the assumptions which led to model (3) do not hold since, for instance, the noise in not signal-independent. Thus only very simple models are useful. According to our experience, the simple choice

$$r_y(k,\ l)=\rho^{|k|+|l|} \qquad 0<\rho<1 \tag{13}$$

for the normalized covariance function turned out to be adequate in most of the cases.

The parameter ρ was estimated by a least-squares fitting of equation (13) to the first lags of the sample covariance function. The noise intensity θ could not be determined experimentally since experimental conditions changed rapidly. Hence θ was chosen *a posteriori*.

With the choice of equation (13) we get a state-space model (3)

$$y(i,\ j)=x(i,\ j) \qquad x(i,\ j+1)=\rho x(i,\ j)+w(i,\ j) \tag{14}$$

where

$$P(k)=\rho^{|k|} \quad \text{and} \quad Q(k)=(1-\rho^2)\rho^{|k|}$$

Note that model (14) is the simplest state-space model. Since the two-sided z-transform of the sequence $P(k)$ is

$$\rho(z)=\sum_{i=-\infty}^{+\infty} \rho^{|i|}\zeta^{i}=\frac{1-\rho^2}{(1-\rho z)(1-\rho z^{-1})}$$

the algebraic Riccati equation (8b) becomes

$$\pi^2(z)+\pi(z)(1-\rho^2)(\theta-\rho(z))-(1-\rho^2)\theta\rho(z)=0$$

Hence $\pi(z)$ can be computed analytically solving a second-order algebraic equation and only an inverse FFT algorithm is necessary to compute the Kalman gain $\boldsymbol{K}$. Moreover, in this case, since the state in (14) has dimension 1, the defining sequence of $\boldsymbol{K}$ has scalar components and is symmetric. Hence one realization is necessary in (9). A good approximation was obtained employing realizations of dimension 2 or 3, i.e. choosing m equal to 4 or 6.

As reported in Section 2, space invariance assumptions lead to a reduction of resolution and to the smoothing of edges, since the covariance function is estimaed over the whole set of points of the image. On the other hand, if we reject this assumption, many of the previous results no longer hold. A way to circumvent this difficulty is to divide the image into vertical strips and estimate for each strip a different ρ, hence a different gain $\boldsymbol{K}$. In this way we get a nonstationary filtering algorithm without increasing numerical complexity.

In our opinion it would be interesting to determine an automatic procedure for partitioning the images or, alternatively, to estimate ρ in a adaptive way during the filtering procedure. Another point to be further investigated is the possibility of partitioning the observed image in nonrectangular subregions.

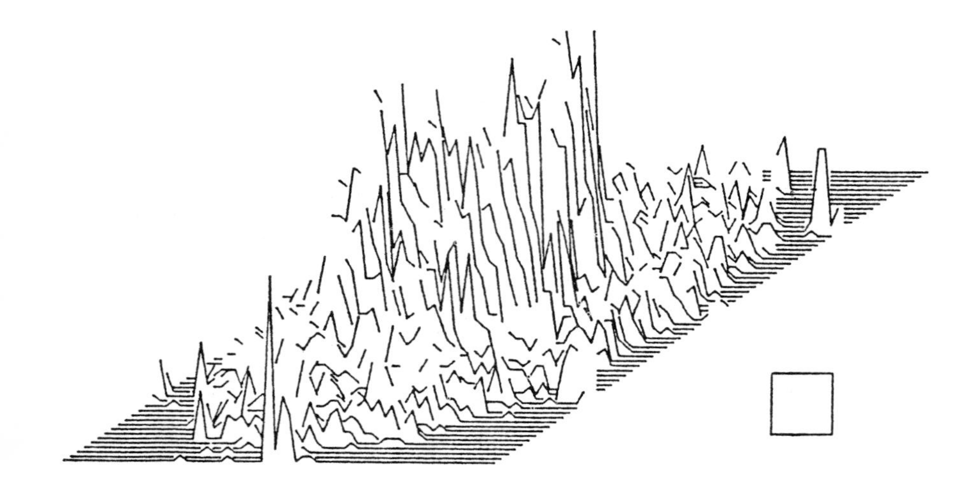

Fig. 1. Scintigraph of human kidneys.

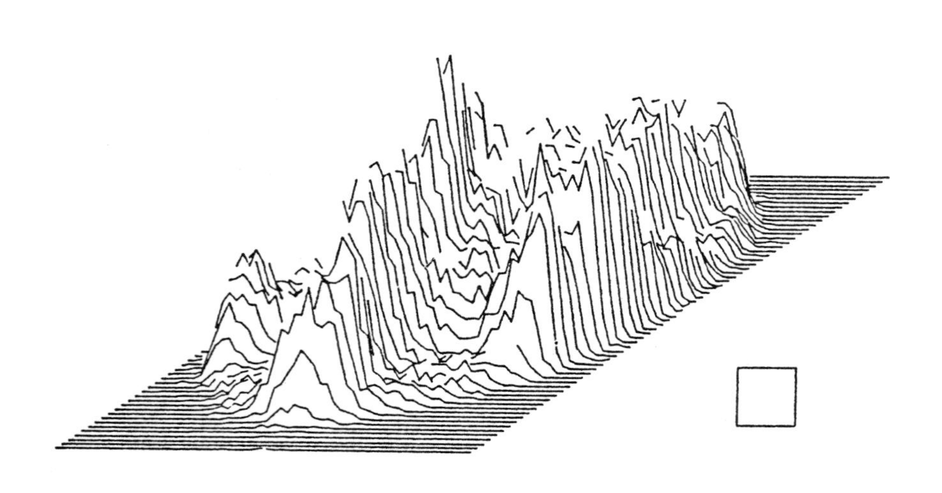

Fig. 2. Front thorax scintigraph.

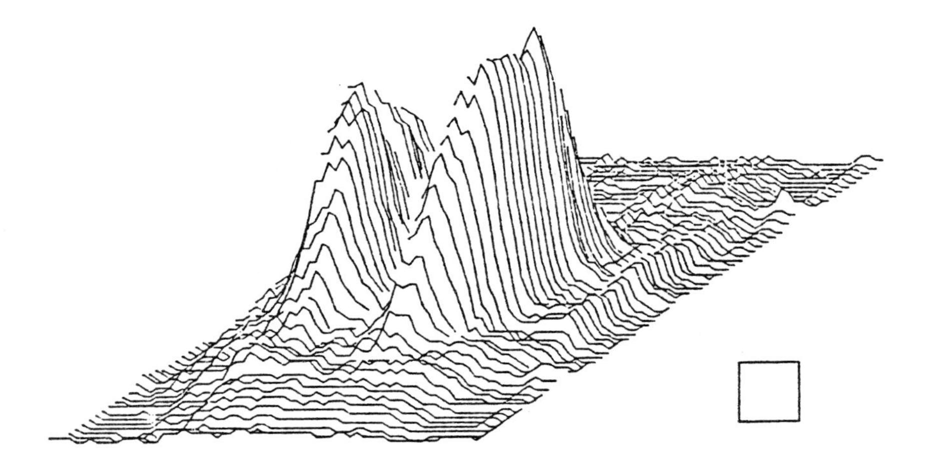

Fig. 3. Image of Fig. 1 filtered in both directions. $\rho = 0{\cdot}96$, S/N = 3.

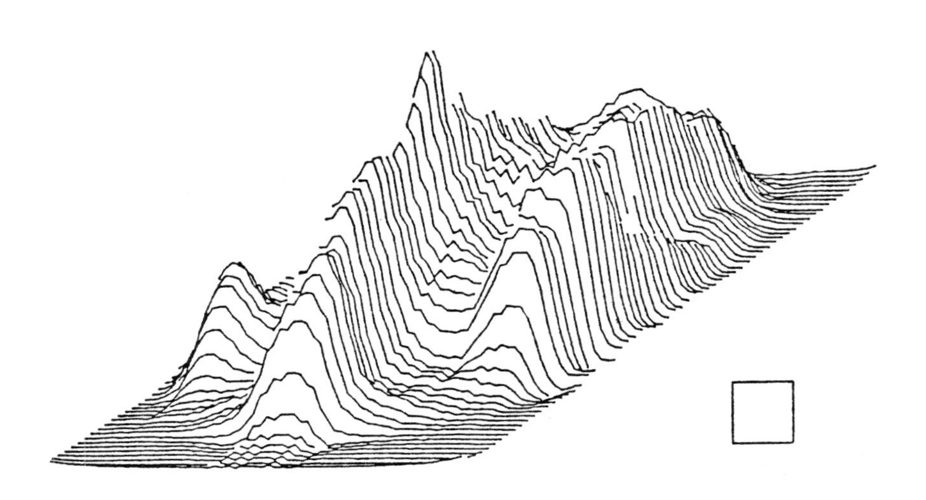

Fig. 4. Image of Fig. 2 filtered in both directions. $\rho = 0{\cdot}97$, S/N = 7.

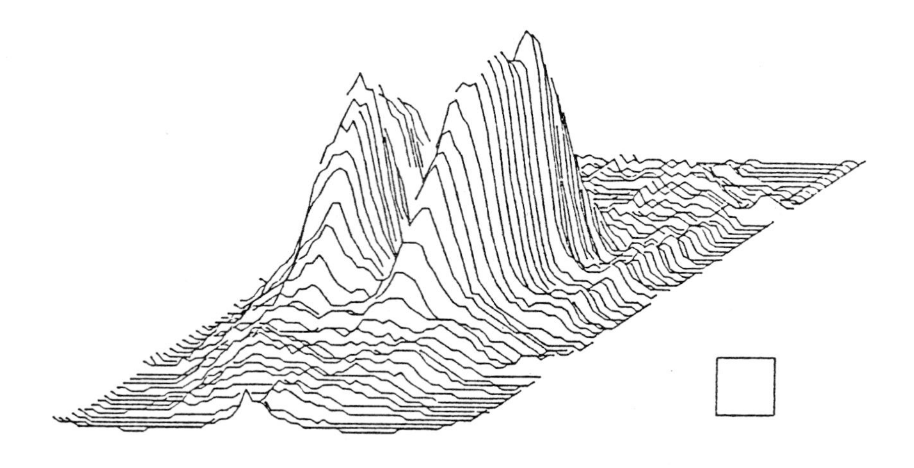

Fig. 5. Image of Fig. 1 filtered in one direction. $\rho = 0{\cdot}96$, S/N = 3.

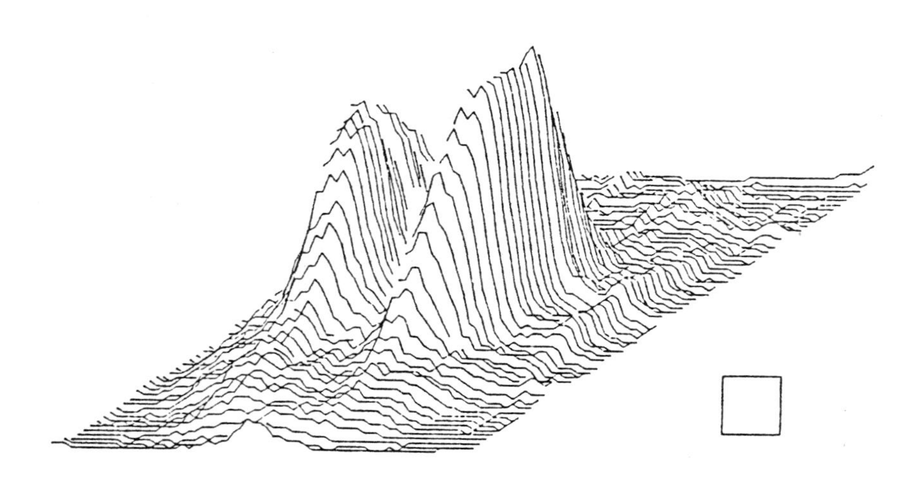

Fig. 6. Image of Fig. 1 filtered in both directions by strip-processing. $\rho_1 = 0{\cdot}99$, $\rho_2 = 0{\cdot}92$, $\rho_3 = 0{\cdot}99$, S/N = 3.

Finally, some results of this filtering procedure are reported in Figs. 1–6. All scintigraphic images are represented in a pseudoperspective view, in order to show more detail, though this is not the display adopted in medical analysis. Figures 1 and 2 show two scintigraphic images representing, respectively, the kidneys and the thorax. The second image has an higher signal-to-noise ratio due to the different radioactive tracer and different concentration. The corresponding processed images are reported in Figs. 3 and 4. They were obtained by an optimal smoother. The first image was also processed by filtering in one direction (from the first to the last row) (Fig. 5) and by a strip-filtering techniques (Fig. 6). In this case the image was divided into three strips, two of them containing essentially only the background. In all cases only the asymptotic gain was employed. For each processed image the parameter ρ and the estimate of the S/N ratio is reported. The value $m=6$ was adopted.

Table 1 lists the computing times, referring to a HP 2100A minicomputer, for different values of m and for various possible alternatives. They are indicative of the complexity of the various algorithms proposed.

From a comparison of Figs. 3 and 4 it turns out that better results are obtained with low S/N ratios, while the improvement of the optimal smoother over a half-plane recursive filter is questionable, if their computing times are taken into account. A little reduction of noise magnitude and a better resolution come from adopting a strip-processing technique, though this particular example could be misleading.

The processed images were quite sensitive to estimated parameters. Hence, even model (13) is a crude approximation; more complex models are of little use, if their parameters must be estimated from the observed image. In order to get better results a deeper analysis of the statistical properties of scintigraphic images, together with more experimental work, is necessary. Moreover, at the present time, it is not clear whether noise reduction or resolution improvement are more useful to medical diagnosis.

Despite these limitations, we claim that Kalman filtering techniques can play an important role in this and other restoration problems and compare favourably with other well established methods.

TABLE 1. Computing times of various algorithms.

	Stationary filter		Nonstat. filter
	One direction	Both directions	both directions
$m=2$	30″	1′ 30″	3′ 00″
$m=4$	50″	2′ 40″	4′ 15″
$m=6$	1′ 10″	3′ 30″	5′ 10″
$m=8$	1′ 50″	5′ 00″	7′ 00″

5. Conclusions

Problems connected with recursive estimation of 2-D processes have been discussed and a Kalman filtering algorithm was developed. Its efficient implementation is achieved by computing the Kalman gains in the z-transform domain and by evaluating convolutions through recursive equations.

This algorithm was employed for the restoration of scintigraphic images. Results obtained so far indicate that for low S/N ratios these techniques are useful and that further studies are worthwhile. In particular further work should examine more closely the statistical properties of the image and the advantages of adopting more complex state-space models. Moreover an on-line adaptive estimation of model parameters would be quite useful for routine image analysis. Finally a comparison with frequency-domain filters seems of interest.

Acknowledgments

We thank Professor V. Cappellini and Professor E. Mosca for helpful suggestions and discussions.

References

1. A. S. Willsky, "Relationships Between Digital Signal Processing and Control and Estimation Theory", *Proc. IEEE*, **66**, No. 9, pp. 996–1017 (Sept. 1978).
2. H. C. Andrews and B. R. Hunt, "Digital Image Restoration", Englewood Cliffs, NJ: Prentice-Hall, 1977.
3. N. K. Bose, "Problems and Progress in Multidimensional System Theory", *Proc. IEEE*, **65**, No. 6, pp. 327–344 (1976).
4. S. Attasi, "Modelling and Recursive Estimation for Double Indexed Sequences". In "System Identification: Advances and Case Studies" (R. K. Mehra and D. G. Lainiotis, eds.). New York: Academic Press, 1976.
5. L. Baroncelli, A. Del Bimbo and G. Zappa, "2-D Kalman Filtering with Application to Biomedical Images", *Tech. Rep. CNR/GNAS* 1/78, Dep. of Elect. Eng., University of Florence, Florence, Italy. (In italian).
6. A. K. Jain and J. R. Jain, "Partial Differential Equations and Finite Difference Methods in Image Processing – Part II: Image Restoration", *IEEE Trans. on Automation and Control*, **AC-23**, pp. 817–834 (Oct. 1978).
7. M. S. Murphy and L. M. Silverman, "Image Model Representation and Line-by-Line Recursive Restoration", *IEEE Trans. on Automation and Control* **AC-23**, pp. 809–816 (Oct. 1978).
8. H. Widom, "Toeplitz Matrices", In "Studies in Real and Complex Analysis" (I. I. Hirschmann, Jr., ed.). Englewood Cliffs, NJ: Prentice-Hall, 1965.
9. F. C. Schoute, M. F. Ter Horst and J. C. Willems, "Hierarchic Recursive Image Enhancement", *IEEE Trans. Circuits and Systems*, **CAS-24**, No. 2, pp. 67–78 (Feb. 1977).

Aspects of 3-D Reconstruction by Fourier Techniques

HANS KNUTSSON, PAUL EDHOLM AND GÖSTA GRANLUND

Department of Electrical Engineering, Linköping University, Linköping, Sweden

1. Introduction

The work presented is an attempt to answer the question: If data are recorded according to the projection procedure illustrated in Fig. 1, what operations should be performed in order to obtain the best possible reconstruction of the object? In a conventional technique for tomography, an approximation of any slice perpendicular to the rotation axis is obtained simply by summing the projections, each one with a given radial displacement.[1] To gain deeper insight into the reconstruction problem its implications in the Fourier domain are studied. This study shows the possibilities and the limitations of a reconstruction method using data obtained according to Fig. 1. Furthermore it shows that these data are sufficient to obtain a reconstruction that in several regards is superior to one obtained by summation.

2. Projection–slice theorem

A very useful theorem, when trying to solve the problem in the Fourier domain, is the projection-slice theorem.[2] This states that the Fourier transform of a projection is a slice of the Fourier transform of the projected object. As shown in Fig. 2 the slice contains the origin and is at right angles to the projecting rays. Fig. 3 shows the positions of the sample planes obtainable when the projection procedure of Fig. 1 is used and $N=8$. If the entire Fourier transform of the object was available, the inverse transform would yield the object. Accordingly we wish to find an interpolation function to calculate the missing values. Figs. 3 and 4 indicate that, as

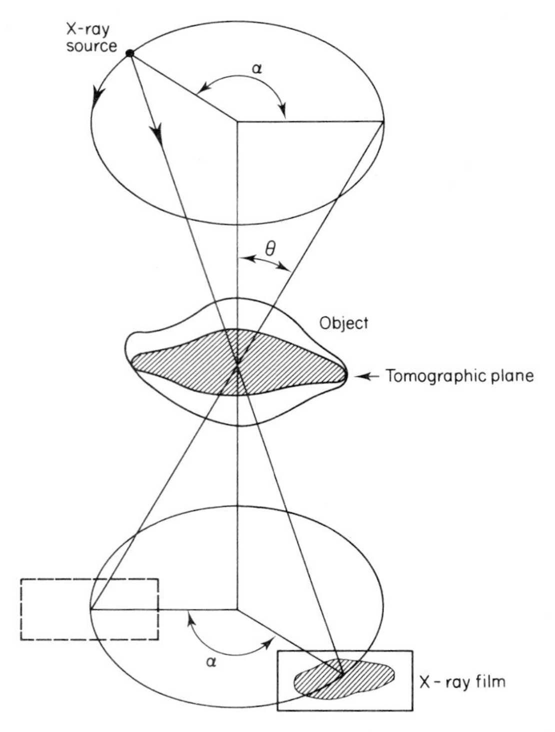

Fig. 1. Tomographic projection procedure. Radiographs taken at angles $\alpha = 2n\pi/N$; $n = 1, 2 \ldots N$.

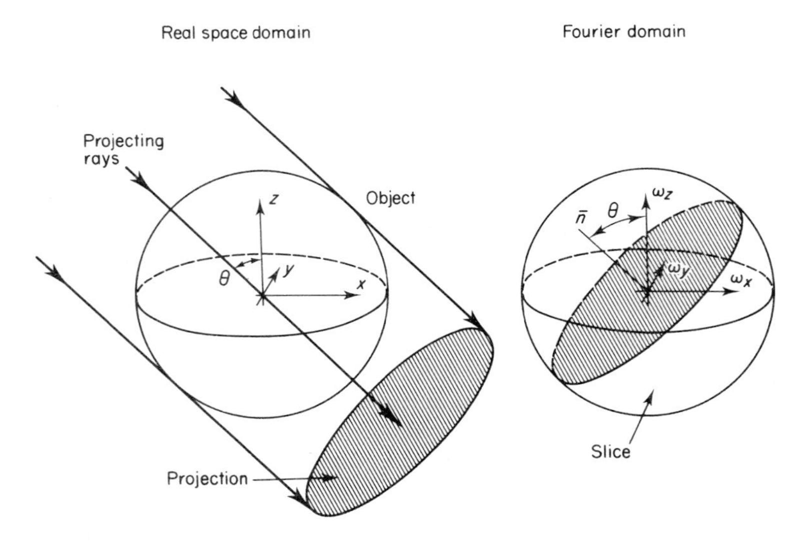

Fig. 2. Projection-slice theorem.

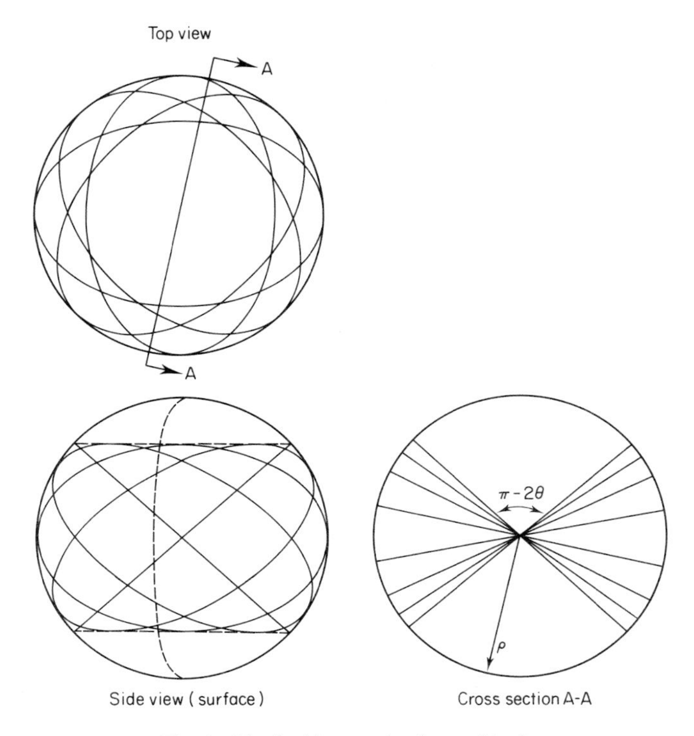

Fig. 3. Obtainable sample planes, $N = 8$.

a result of the projection procedure, there will be an empty cone in the Fourier transform space within which no values of the transform can be calculated. Outside the cone the transform can always be determined with sufficiently large N for finite frequencies. An attempt to obtain an exact reconstruction of the Fourier transform, in areas where this is possible, would however, be very difficult and time consuming due to the highly irregular sampling pattern. In order to simplify the problem we restrict the operations performed on data in the Fourier domain to weighting operations.

3. Three-dimensional sample density

To find an appropriate weighting function our basic assumption is that with no *a priori* knowledge of the object, all parts of the Fourier transform can be considered to contain the same amount of information per unit volume. Consequently we would like the sample density to be constant in the Fourier domain. The relative sample density D_{s3} is defined as N^{-1} times the

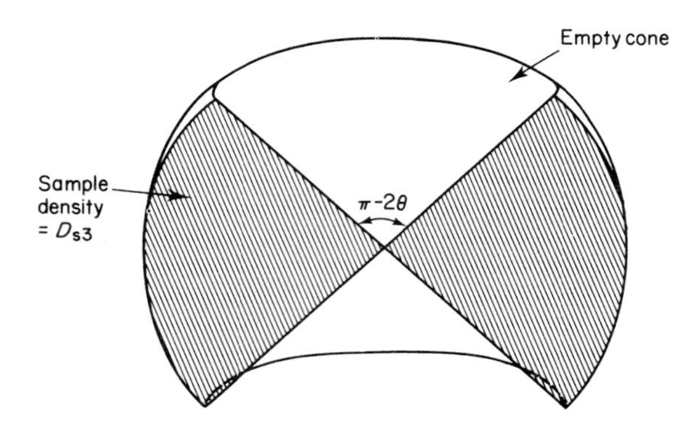

Fig. 4. Sample density in the Fourier domain.

sample plane area per unit volume in the limit $N \rightarrow \infty$. It can be shown that

$$D_{s3} \sim \rho^{-1} |\cos \phi|^{-1} = |\omega_1|^{-1}$$

where ρ, ϕ, ω_1 are coordinates for points on one of the sample planes, as illustrated in Fig. 5. From symmetry the expression is necessarily valid for all sample planes. An intuitive understanding of this result is gained by studying the distance between two closely spaced sample planes as shown in Fig. 5. The distance is found to be approximately

$$d \sim |\omega_1| = D_{s3}^{-1}$$

Considering the discussion above the obvious choice of weighting function is

$$f_{w3} = D_{s3}^{-1} = |\omega_1|$$

since this function will make the weighted sample density constant in all parts of the Fourier transform outside the empty cone.

The next step is to project the Fourier transform onto a plane in the way shown in Fig. 6. The reason is that the result of a reconstruction will have to be displayed as a two-dimensional function; i.e. only a slice of the object can be displayed at a time. Use of the projection-slice theorem backwards

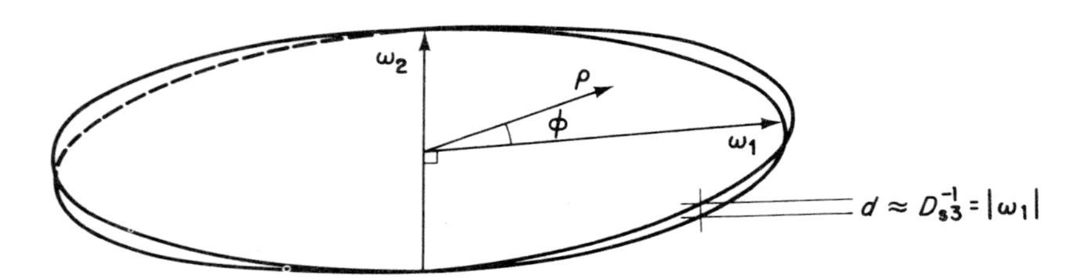

Fig. 5. Two sample planes.

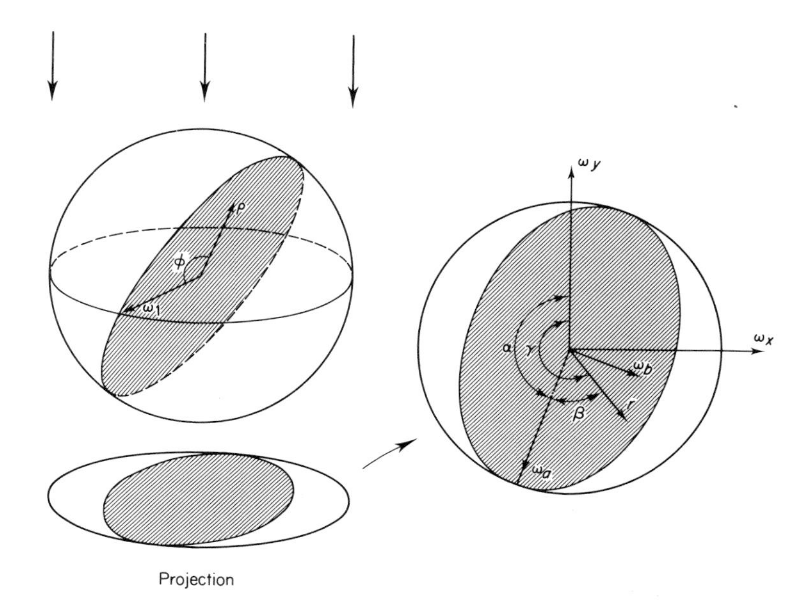

Fig. 6. Projection of sample plane.

implies that the corresponding operation in the Fourier domain is projection.

We now conclude that it is possible to find two-dimensional weighting functions that, when applied individually to the projections of the Fourier sample planes, i.e. the Fourier transform of each object projection, have an effect equivalent to that of the three-dimensional weighting functions. We need only consider the coordinate transformations caused by the projection in Fig. 6. The two-dimensional weighting function becomes

$$f_{w2} = |\omega_a| = r|\cos\,\beta|$$

If the fixed coordinate system is used, f_{w2} can be expressed as

$$f_{w2} = r|\cos\,(\gamma - \alpha)|$$

We can see that f_{w2} is of the same form for all projections except that it is rotated with an angle α.

4. Two-dimensional sample density

Another desired property of the weighting function is that, in order to produce an object reconstruction without frequency distortion, it should make the weighted two-dimensional sample density constant:

$$\sum_{\text{all}} h_{w2} = C$$

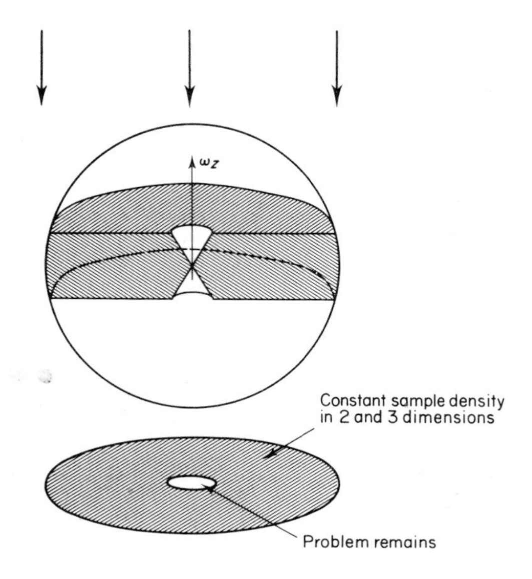

Fig. 7. Cut-off in ω_z direction.

But

$$\sum_{\text{all}} f_{w2} \approx C_1 r$$

and we can see that f_{w2} does not have this property, the reason for this is obvious if Fig. 4 is studied, f_{w2} makes the three-dimensional sample density constant and because of the empty cone the projected density will be proportional to r.

A practical consideration enters the discussion here. When reconstructing a slice of a human body it is in most cases not necessary or even desired that this slice is very thin. This implies that, as the slice thickness is dependent on the resolution in the z-direction of the object reconstruction and this corresponds to some cutoff frequency ω_{zc} in the Fourier domain,

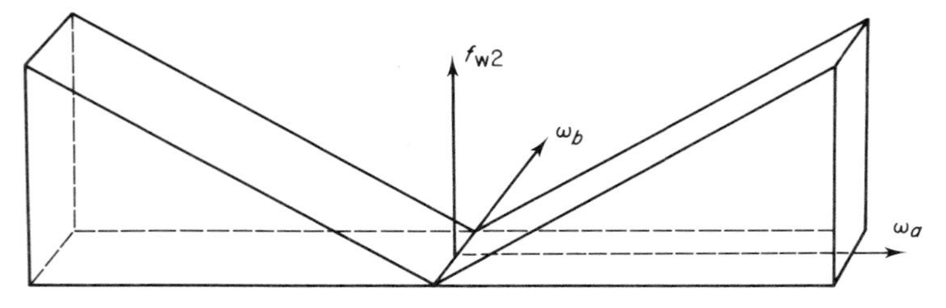

Fig. 8. Principal form of resulting two-dimensional weighting function.

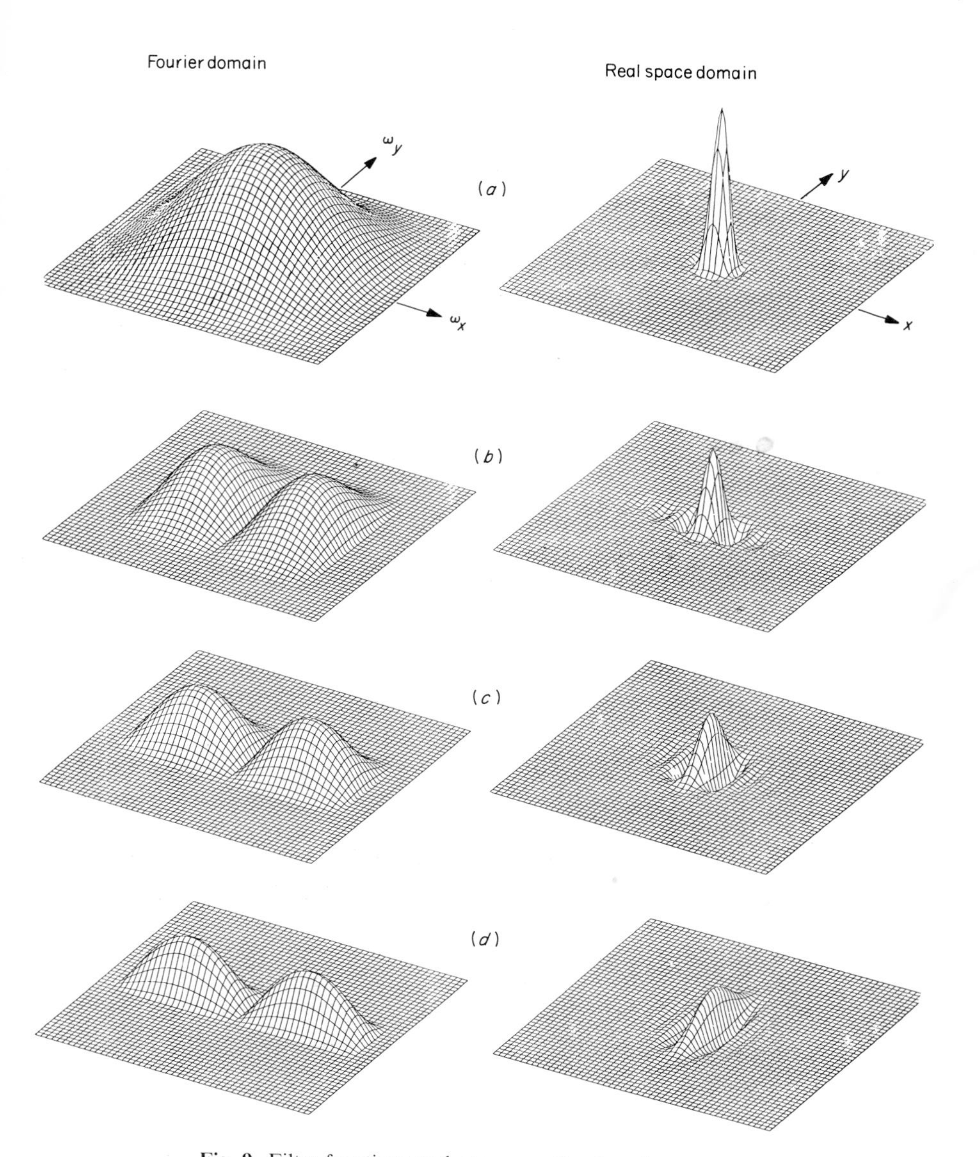

Fig. 9. Filter functions and corresponding impulse responses.

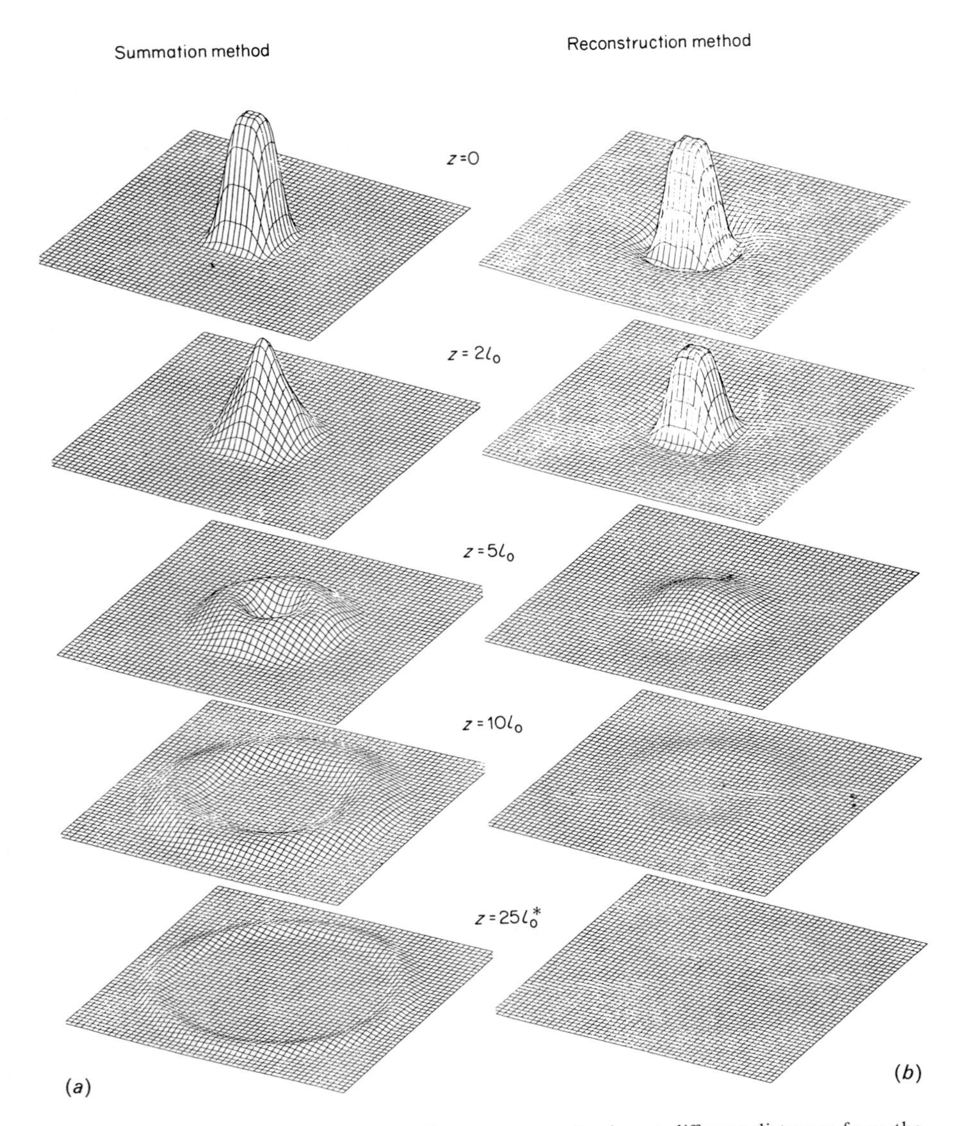

Fig. 10. Contribution from a small radiopaque square lamina at different distances from the slice centre.

f_{w3} can be set to zero for $\omega_2 > \omega_{zc}$. In Fig. 7. The effect of this cut off can be studied and it is found that the two-dimensional as well as the three-dimensional sample density can be made constant if

$$\sqrt{\omega_x^2+\omega_y^2} \geqq \omega_{xc}/\tan\,\theta$$

But for

$$\sqrt{\omega_x^2+\omega_y^2} < \omega_{zc}/\tan\,\theta$$

this is not possible. The resulting two-dimensional weighting function f_{w2} after cutoff of f_{w3} is shown in Fig. 8.

5. Simulation

To gain further insight into the behaviour of this reconstruction method a number of simulations has been performed. The form of the weighting function we have chosen and its impulse response is outlined in Fig. 9(*b*). The smoothed form of the weighting function is employed to obtain an impulse response with less ringing. For this reason the weighting function used when simulating the summation method is a raised cosine window as shown in Fig. 9(*a*). Figures 10(*a*) and (*b*) show the contribution to the slice reconstruction from a small square lamina, at five different distances from the slice centre, for the summation and reconstruction method respectively. Figure 10 demonstrates that the reconstruction method features both lower distortion of objects near the slice centre and a much more effective cancellation of more distant objects.

References

1. R. N. Bracewell, "Strip Integration in Radio Astronomy", *Aust. J. Phys.*, No. 9, p. 198 (1956).
2. P. Edholm, "The Tomogram. Its Information and Content." *Acta Radiol. Suppl.* No. 193 (1960).
3. P. Edholm and L. Quiding, "Elimination of Blur in Linear Tomography." *Acta Radiol. Diagnosis* **10**, p. 441 (1970).
4. D. G. Grant, "Tomosynthesis. A Three-Dimensional Radiographic Imaging Technique." *IEEE Trans. Biomed. Eng.*, **BME-19**, p. 20 (1972).
5. T. M. Peters, "Image Reconstruction from Projections." Electrical Eng., report from Univ. of Canterbury, Christchurch, New Zealand, No. 16, June 1973.
6. R. M. Mersereau and A. V. Oppenheim, "Digital Reconstructions of Multi-dimensional Signals from their Projections," *Proc. IEEE*, No. 10 (1974).
7. P. Edholm, G. Granlund, H. Knutsson and C. Petersson, "Ectomography. A New Radiographic Method for Reproducing a Selected Layer of Varying Thickness," to be published.
8. C. Petersson, P. Edholm, G. Grandlund and H. Knutsson, Ectomography. A New Radiographic Reconstruction Method. Computer Simulated Experiments," to be published.

Design of a Computerized Emission Tomographic System

S. BELLINI

Centro di Studio per le Telecomunicazioni Spaziali del C.N.R., Italy

C. CAFFORIO

Istituto di Elettrotecnica ed Elettronica del Politecnico di Milano, Milan, Italy

M. PIACENTINI

Centro di Studio per le Telecomunicazioni Spaziali del C.N.R., Italy

F. ROCCA

Istituto di Elettrotecnica ed Elettronica del Politecnico di Milano, Milan, Italy

1. Introduction

The application of tomographic reconstruction algorithms to radiological surveys has been the paramount event in medical diagnostics during the last few years. Even if possible in principle, the extension of these methods to nuclear medicine has not yet attained similar success. Lack of resolution and tissue absorption are the chief problems, but not the only ones. On the other hand emission tomography, in contrast with its X-ray counterpart, easily obtains multiple slices and could allow metabolic studies. Leaving to others the task of devising new radiopharmaceuticals, the engineer has to supply the need for very fast image reconstruction algorithms.

In this paper we present the results of a study carried out in order to construct an emission tomography system. We have recently developed a new algorithm for tissue absorption compensation. A brief summary of the algorithm, without mathematical details, is presented here.

2. Back-projection algorithm

The problem in tomography is to determine a function of two variables $a(x, y)$ from a set of projections, i.e. side views $p(r, \phi)$ taken at angular values

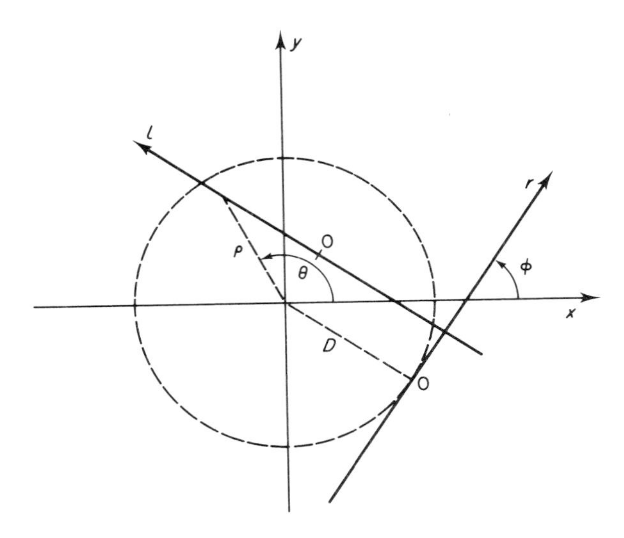

Fig. 1. Projection geometry.

of ϕ (see Fig. 1):

$$p(r, \phi)=\int_{-\infty}^{+\infty} a(x, y)\mathrm{d}l \qquad l{:}x \cos \phi+y \sin \phi-r=0$$

It is well known[1-4] that

$$P(R, \phi)=A(R, \phi) \tag{1}$$

where $P(R, \phi)$ is the monodimensional Fourier transform of $p(r, \phi)$ and $A(R, \phi)$ is the two-dimensional Fourier transform, in polar coordinates, of $a(x, y)$.

If $P(R, \phi)$ is known for every ϕ then $A(R, \phi)$ is known and $a(x, y)$ can be recovered by inverse Fourier transform. In polar coordinates we have

$$\begin{aligned} a(\rho, \theta) &= \frac{1}{2\pi} \int_0^{\pi} \frac{1}{2\pi} \int_{-\infty}^{+\infty} P(R, \phi)\ |R|\ \exp(\mathrm{j}R\rho \cos(\theta-\phi))\, \mathrm{d}R\mathrm{d}\phi \\ &= \frac{1}{2\pi} \int_0^{\pi} p_F(\rho \cos(\theta-\phi), \phi)\, \mathrm{d}\phi \end{aligned} \tag{2}$$

where $p_F(r, \phi)$ is the original projection filtered by a $|R|$ filter. Equation (2) states that the function $a(\rho, \theta)$ is obtained by back-projecting each filtered projection. By "back-projecting" we mean to add the value $p_F(r, \theta)$ to all the image points lying on the line passing through r and drawn perpendicularly to the projection.

By repeated application of the algorithm, several contiguous slices can be processed. In this way a complete three-dimensional description of radionuclide distribution is obtained. In our project a γ-camera is used as a detector and 64 projections are available.

The stringent requirement for a very fast algorithm necessitates various simplifications. Low resolution and signal-to-noise ratio allow many simplifications without appreciable loss of image quality. As a consequence of γ-camera nonuniformity, the signal-to-noise ratio is never better than 20 dB in the well behaved zones of the image. The average signal-to-noise ratio can be even lower.

Let us now analyse all the choices and trade-offs arising in the system design.

2.1. Projection number

This fundamental parameter is strictly connected with image quality and algorithm speed. Some authors[5-6] assert the need of 200 projections over 360° in order to get accurate reconstructions of 64 × 64 pixel images. In spite of this it is a common practice to use only 60 or 30 views in nuclear medicine applications.

Some analytical insight can be gained into the problem. It has already been noted that $P(R, \phi) = A(R, \phi)$. Therefore taking N projections over 180° corresponds to knowing $A(R, \phi)$ along straight lines angularly spaced by π/N. For large values of R angular sampling becomes inaccurate. Let us consider an impulse located at (ρ_0, θ_0):

$$a(\rho, \theta) = \delta(\rho - \rho_0, \theta - \theta_0)$$

Its Fourier transform can be expanded by means of Bessel functions

$$A(R, \phi) = \exp(-jR\rho_0 \cos(\theta_0 - \phi)) = \sum_{-\infty}^{+\infty} J_n(R\rho_0)\exp(-jn(\theta_0 - \phi + \pi/2))$$

For a given $R\rho_0$, $A(R, \phi)$ is 2π-periodic and is sampled at a sampling rate of π/N. To avoid aliasing, the terms $J_n(R\rho_0)$ have to be very small for $n > N$. The worst case is attained when both R and ρ_0 reach their maximum values. If d is the projection sampling interval and M is the number of samples per projection we have

$$\max(R\rho_0) = 2\pi \frac{1}{2d} d \frac{M}{2}$$

In order to have $J_n(x) < 0{\cdot}01$ the condition $n \geqq x + 5$ must be verified; about 105 views are then needed if $M = 64$.

However the detector assembly, i.e. γ-camera and collimators, resolves only 15 mm at a distance of 20–30 cm. This means that the maximum signal frequency is less than the Nyquist frequency connected with projection sampling. On the other hand, it is uncommon for typical images to need full resolution at a distance from the centre greater than 20*d*. Taking these facts into account gives $(R\rho_0)_{max}=25$–30 and so we use 32 projections.

Angular subsampling creates artifacts with a radial structure mainly at the image borders. Measured amplitudes of these artifacts range over a few per cent of the image maximum value. Image quality does not suffer much from these effects.

2.2. Filtering

The "filtered back-projection" algorithm requires a $|R|$ filter. In effect such a frequency response would enhance the out-of-band noise. For this reason the $|R|$ filter has been combined with a low-pass filter. Cut-off frequency is not a critical parameter; it is more important not to have a sharp cutoff in order to avoid ringing.

We use a raised cosine transition low-pass filter with a -3dB frequency slightly lower than half the Nyquist frequency. With typical images no loss of resolution is experienced. Filtering is done by direct convolution and satisfactory results have been obtained by the use of a 13-point impulse response. Figure 2 shows it actual frequency response.

We have compared the images obtained using the convolutional filter with the more accurate ones obtained when filtering is accomplished by means of FFT techniques. Degradation is noticeable but not enough to justify the waste of time associated with the second technique.

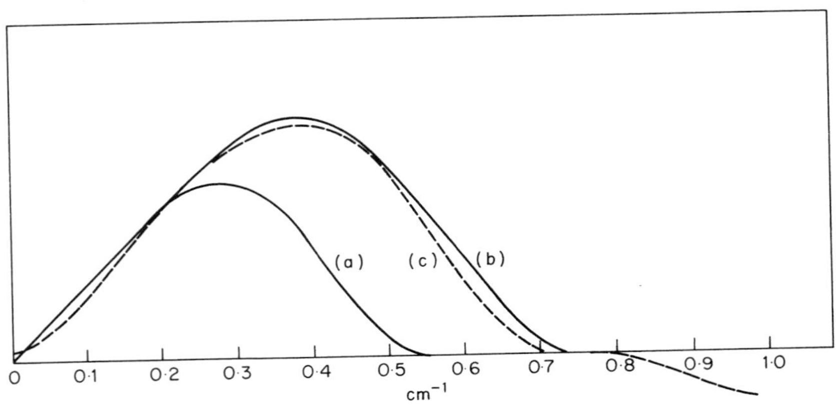

Fig. 2. Frequency response curves: curve a: sampling step = 0·68 cm; curve b: sampling step = 0·5 cm; curve c: approximation of b with a 13-point FIR filter.

2.3. Back-projection

The value of the reconstructed matrix in a point (i, j) is given by the sum of the values that each filtered projection assumes at the coordinate where the point itself is projected. Input data are known at discrete points; therefore some form of interpolation is needed. As a crude approximation we simply take the value of the filtered projection at the nearest integer coordinate. Thus a lot of computing time is saved. Moreover it becomes feasible to use pre-calculated cross-reference tables which give for each sample of every projection the points in the reconstructed image where its value has to be added.

Unlike filtering, which is performed in floating point notation, back-projection has to be carried out in integer notation because of core memory size. We evaluate the errors introduced by such a procedure as follows.

Let the values of the image elements be uniformly distributed in the interval 0 to L. Their mean-square value is $L^2/3$. The use of integer numbers is equivalent to the use of a quantizer with step $\Delta = 1$. The quantization noise is uniformly distributed between the values $-\frac{1}{2}$ and $+\frac{1}{2}$. Then the signal to noise ratio is given by

$$\mathrm{S/N} = 10 \log_{10}\left(\frac{L^2}{3}\frac{12}{\Delta^2}\right) = 6 + 20 \log_{10} L$$

Typical values for L are in the range of a few tens. They can be increased by means of a proper choice of filter weights and therefore noise due to the use of integer numbers is not troublesome.

The noise introduced by lack of interpolation has worse effects and some reasoning is required to get an estimate of its value. The use of the nearest known value instead of the correctly interpolated one can be modelled as a jitter of the projections. Let us suppose that the jitter is uniformly distributed in the range $-\frac{1}{2}$ to $+\frac{1}{2}$.

Let $p(r)$ be any given projection (we drop here the angular variable ϕ) and let the image be an impulse. Then

$$p(r) = \sum_{i=1}^{M} p_i(r - r_{0i}) \qquad P(R) = \sum_{i=1}^{M} p_i \exp(-\mathrm{j}Rr_{0i})$$

The jitter can be taken into account by

$$r_i = r_{0i} + \Delta_i$$

The spectrum of the jittered projection is then

$$\hat{P}(R) = \sum_{i=1}^{M} p_i \exp(-\mathrm{j}Rr_{0i}) \exp(-\mathrm{j}R\Delta_i)$$

For small $R\Delta_i$, we can approximate $\exp(-jR\Delta_i)$ with the first two terms of its series expansion:

$$\hat{P}(R)=\sum_{i=1}^{M} p_i \exp(-jR\Delta_i)=P(R)-\sum_{i=1}^{M} p_i \exp(-jRr_{0i})jR\Delta_i$$

The noise power spectrum is then

$$|N(R)|^2=R^2|P(R)|^2\Delta_i^2=|P(R)|^2R^2/12$$

We must now take into account the fact that in order to determine the value of each picture element 32 projections have to be averaged. This reduces the noise by 15 dB, considering the disturbances caused by the jitter in different projections as independent. However, this is only a crude approximation; in fact, if in consecutive projections the same picture element is projected in different positions of the same interval between samples, the errors will tend to compensate. This improves the previously considered margin, and gives a signal-to-noise ratio of about 16 dB at the Nyquist frequency. At the maximum frequency of typical images the signal-to-noise ratio is about 30dB.

3. Compensation of tissue absorption

Several methods have been proposed to compensate tissue absorption in emission tomography, but only those based on iterative techniques and on an *a priori* knowledge of the absorbing body's structure may be considered accurate. The main drawback is the very long computing time that they require.

If the attenuation per unit length can be considered constant throughout the absorbing body in which the emitting sources are embedded, it is possible to obtain an analytical solution to the problem provided that the countour of the body is convex and that we know it.

The projection taken at an angle ϕ can be expressed as

$$p(r, \phi, \alpha)=\int_{-\infty}^{+\infty} a(\rho, \theta) \exp(-\alpha(D(r, \phi)+l))\mathrm{d}l \tag{3}$$

The factor $\exp(-\alpha D(r, \phi))$ can be compensated, so that we shall ignore it from here on. Note that without attenuation $p(r, \phi+\pi)=p(-r, \phi)$, whereas in the presence of an absorbing body no simple relation holds between two antipodal projections. Equation (3) can be rewritten as

$$p(r, \phi, \alpha)=\int_{-\infty}^{+\infty} a(\rho, \theta) \exp(-\alpha\rho \sin(\theta-\phi))\mathrm{d}l$$

Its Fourier transform is

$$P(R, \phi, \alpha)=\int_{-\infty}^{+\infty}\int_{-\infty}^{+\infty} a(\rho, \theta) \exp(-\alpha\rho \sin(\theta-\phi)) \exp(-jRr)\mathrm{d}l\mathrm{d}r$$

This double integral can be written in polar coordinates as

$$P(R, \phi, \alpha)=\int_{0}^{2\pi}\int_{0}^{\infty} a(\rho, \theta) \exp(-\alpha\rho \sin(\theta-\phi)-jR\rho \cos(\theta-\phi))\rho\mathrm{d}\rho\mathrm{d}\theta$$

If $\alpha=0$ this reduces to equation (1).

The problem is then to find $A(R, \phi)=P(R, \phi, 0)$ from $P(R, \phi, \alpha)$. It can easily be verified that the function of the three variables $P(R, \phi, \alpha)$ satisfies the partial differential equation

$$\frac{\partial P}{\partial \alpha}=-\frac{\alpha}{R}\frac{\partial P}{\partial R}-\frac{j}{R}\frac{\partial P}{\partial \phi}$$

Solving this equation and setting the appropriate boundary conditions one has the fundamental formula

$$P(R, \phi, 0)=P((R^2+\alpha^2)^{1/2}, \phi+jSh^{-1}(\alpha/R), \alpha)$$

The evaluation of $P(R, \phi, \alpha)$ at radial frequencies $(R^2+\alpha^2)^{1/2}$ is easily accomplished by means of interpolation. The calculation of the function for complex values of ϕ is conveniently done by Fourier techniques. The details can be found in ref. 11. It is convenient to work on averaged antipodal projections, i.e. on

$$p_a(r, \phi, \alpha)=\tfrac{1}{2}p(r, \phi, \alpha)+\tfrac{1}{2}p(-r, \phi+\pi, \alpha)$$

The algorithm requires the following steps:

(1) Take the Fourier transforms $P_a(R, \phi, \alpha)$. This can be done by N FFTs.
(2) Calculate $P_a((R^2+\alpha^2)^{1/2}, \phi, \alpha)$ by interpolation. This is a periodic function of ϕ, with period 2π.
(3) Calculate the Fourier coefficients

$$a_k(\mathrm{R})=\frac{1}{2\pi}\int_{0}^{2\pi} P_a((R^2+\alpha^2)^{1/2}, \phi, \alpha) \exp(-jk\phi)\mathrm{d}\phi$$

This is done by FFTs.

(4) Calculate

$$b_k(R)=a_k(R)/Ch(k\ Sh^{-1}(\alpha/R)) \qquad R\neq 0$$

(5) Calculate

$$A(R, \phi)=\sum_{k=-\infty}^{\infty} b_k(R) \exp(jk\phi) \qquad R\neq 0.$$

This is done by FFTs.

(6) If needed, calculate

$$A(0, \phi)=\frac{1}{2\pi}\int_0^{2\pi} P_a(\alpha, \phi, \alpha)\, d\phi$$

(7) Multiply $A(R, \phi)$ by $|R|$, calculate $p_F(r, \phi)$ by N FFTs and back-project.
Figures 3 and 4 give an example of the results that can be obtained.

4. Conclusions

We have reported the design procedures of an ECT-single photon system together with a new algorithm devised to correct tissue absorption. This system, built by SELO S.p.A., after tests at the Nuclear Medicine Department of the Hospital of Niguarda, Milan (Fig. 5) is now commercially available.

Acquisition times, with standard dosages of radiopharmaceuticals range from 10 to 20 minutes. Sixty-four simultaneous slices can be reconstructed at a rate of about 20 seconds per slice. Real-time operating systems allow sharing of acquisition and elaboration times. The attenuation correcting

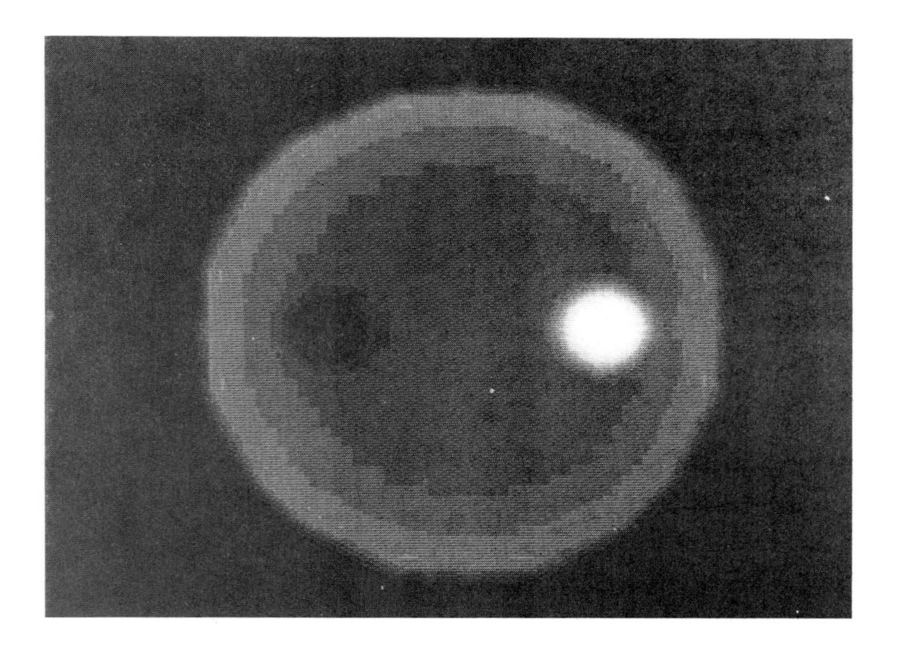

Fig. 3. Image reconstruction from attenuated projections.

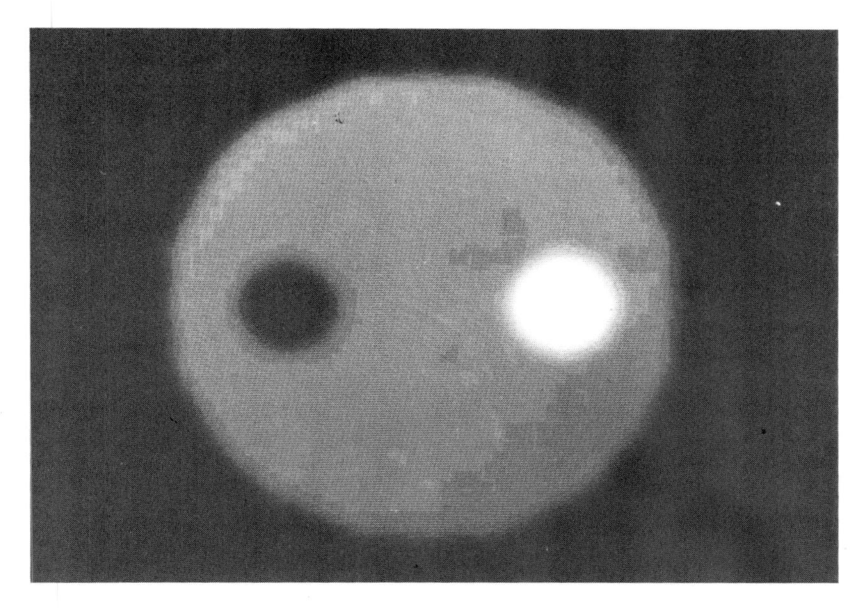

Fig. 4. Image reconstruction with attenuation correction.

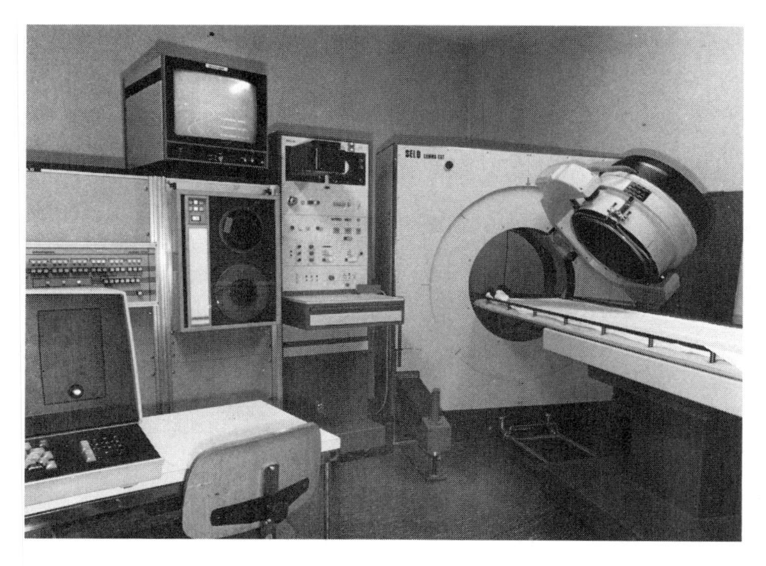

Fig. 5. The emission tomographic system. The minicomputer used is very similar to the VARIAN/620L.

algorithm is rather slower: it takes about three minutes per corrected slice. However we do not consider this a serious drawback. The correction procedure can be considered as a second approximation step, not to be used as a routine procedure. Spare computer time can therefore be devoted to this job. The results obtained are satisfactory, but studies are in progress towards better resolution.

References

1. H. H. Barret and W. Swindell, "Analog reconstruction methods for transaxial tomography", *Proc. IEEE*, **65**, No. 1 (Jan. 1977).
2. R. M. Mersereau and A. V. Oppenheim, "Digital reconstruction of multi-dimensional signals from their projections," *Proc.* IEEE, **62**, No. 10 (Oct. 1974).
3. L. A. Shepp and B. F. Logan, "The Fourier reconstruction of a head section," *IEEE Trans. on Nuclear Science*, **NS-21** (June 1974).
4. T. F. Budinger and G. T. Gullberg, "Three dimensional reconstruction in nuclear medicine by iterative least squares and Fourier transform techniques," *IEEE Trans. on Nuclear Science*, **NS-21**, No. 3 (June 1974).
5. W. I. Keyes, "A practical approach to transverse-section gamma-ray imaging", *British Journal of Radiology*, **49** (1976).
6. T. F. Budinger, S. E. Derenzo, G. T. Gullberg, W. L. Greenberg and R. H. Huesman, "Emission computer assisted tomography with single-photon and positron annihilation photon emitters," *Journal of Computer Assisted Tomography* (1977).
7. W. I. Keyes, R. Chesser and P. E. Undrill, "Transverse emission tomography" 7th LH Gray Conf.: Medical Images.
8. R. J. Jaszczak, P. H. Murphy, D. H. Huard and J. A. Burdine, "Radionuclide emission computed tomography of the head with 99 mTC and a scintillation camera", *Journal of Nuclear Medicine*, **18** (1977).
9. T. F. Budinger and G. T. Gullberg, "3-D reconstruction in nuclear medicine emission imaging", *IEEE Trans. on Nuclear Science*, **NS-21** (June 1974).
10. R. C. Hsieh and W. G. Wee, "On methods of three-dimensional reconstruction from a set of radioisotope scintigrams", *IEEE Trans. on Sys., Man, and Cyb.*, **SMC-6**, No. 12 (Dec. 1976).
11. S. Bellini, C. Cafforio, M. Piacentini and F. Rocca, "Compensation of tissue absorption in emission tomography", *IEEE Trans. on Acoustics Speech and Signal Processing*, **ASSP-27**, No. 3 (June 1978).

4.3. Applications to radar systems

Some Importance Sampling Techniques with Applications to Radar Systems

G. BENELLI, V. CAPPELLINI and E. DEL RE

Istituto di Elettronica, Facoltà di Ingegneria, Florence, Italy

V. FERRAZZUOLO

SMA, Florence, Italy

R. PIERONI†

Istituto di Elettronica, Facoltà di Ingegneria, Florence, Italy

I. SALTINI

SMA, Florence, Italy

Two importance sampling techniques are presented for performing fast digital simulations, greatly reducing the computing time in low-error-probability estimations. The techniques are based on the distortion of noise characteristics at suitable points of the system to be simulated in such a way as to increase the error rates. Such distortion is later compensated by means of a suitable weighting at the system output. The application of these techniques to false alarm probability estimation of a CFAR radar is described. Comparisons with standard Monte Carlo methods are shown, giving good performance agreements in obtained error-probability values and much higher processing rates for the techniques described.

1. Introduction

Using digital computers it is possible to perform simulations of complex systems, evaluating the main performance and efficiency characteristics, without the implementation of high-cost hardware simulators or laboratory testing the actual systems. An important characteristic to be evaluated in

† Present address: SIP, L'Aguile, Italy.

communication and radar systems is represented by the "error probability". Values of the order of 10^{-6}–10^{-12} can readily be found.

Standard Monte Carlo simulation techniques require very considerable computing times—in some cases the simulation is practically impossible. Techniques for reducing the simulation times to practical values are therefore required.

Some techniques on these lines are represented by the so-called fast Monte Carlo simulation or reduced variance methods.[1-3]. These techniques, also called "importance sampling", are essentially based on the strategy of increasing the error sources, and weighting the values obtained at the output to compensate for the previous distortion.

In this paper two techniques of this kind are defined so as to be suitable for simulation of CFAR (constant false alarm rate) radar systems, giving fast, accurate estimates of the false alarm probability. Comparisons with standard techniques are presented to illuminate the efficiency of these importance sampling techniques.

2. Description of the importance sampling techniques

The precision in the evaluation of performances of a system by computer simulation depends on the number of data processed. Nevertheless, in general this precision increases only slowly with the number of data. To illustrate this we consider the evaluation of the mean of a random variable y. The mean is obviously a random variable with Gaussian distribution and standard deviation σ,[4] which is proportional to $(1/n)^{1/2}$. For instance, to obtain $\sigma/2$ it is necessary to process a quadruple number of data.

In many cases the time required to elaborate the samples necessary to obtain a given precision can be very high using the normal Monte Carlo technique. This is particularly true when the parameter to be evaluated has very low probability of occurence and/or the system is complex. In these cases it can be convenient to use a Monte Carlo technique with variance reduction. In methods of this type the input of the system is distorted so that it has on an average a higher number of the desired observations, which determine the behaviour of the system. For example, to evaluate the error probability of a communication or a radar system, the noise level can be increased to have a higher number of errors.

In the following we illustrate briefly the principles of these methods and after we have considered their application to the evaluation of the false alarm probability P_{fa} in a radar system.

Firstly we describe a simple method with variance reduction, called the "stratified sampling technique".[1,2] The cumulative distribution $F(y)$ of the random variable y, which is at the input of a system, is divided over m "strata": the ith stratum extends from $F(y)=\alpha_{i-1}$ to $F(y)=\alpha_i$, with $\sigma=0$

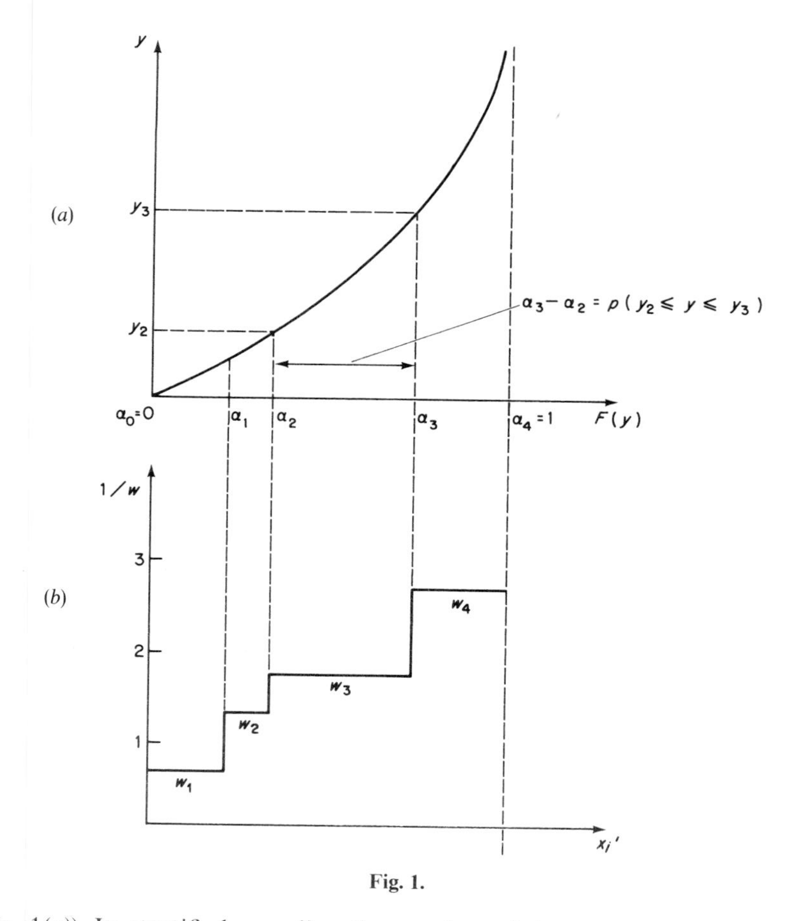

Fig. 1.

(Fig. 1(*a*)). In stratified sampling the number of the observations extracted from some strata is modified with respect to the normal Monte Carlo technique. That is, in these strata which are most important to the result, a higher number of observations are required than in the normal case. In the evaluation of the system performance it is necessary to take account of this distortion at the inputs. This is done through a suitable weighting factor w_i associated to the ith stratum (Fig. 1). For example, in the evaluation of the mean value $\bar{y}$ of a random variable, if we write y'_j for the observation in the stratified sampling, we have

$$\bar{y} = \frac{1}{n} \sum_{j=1}^{n} w_j y'_j \tag{1}$$

n being the number of observations. For the application of this technique it is important to make a suitable choice of the numbers n_i of observations in the ith stratum and of w_i. In general this choice depends on the particular system considered and the desired precision. Nevertheless it is found that a good choice for w_i and n_i is the following:[4,5] the numbers n_i are the nearest integer to the product of the actual probability of an observation in the ith stratum for the standard deviation in the stratum itself. Then the weight factor w_i can be chosen equal to the ratio between the actual probability of an observation from the ith stratum and the distorted probability n_i/n.

The technique of stratified sampling is improved by increasing the number m of the strata. For an m which tends toward infinity, the ratio n_i/n becomes a function of y, which we denote by $p^*(y)\mathrm{d}y$, while the difference $(Q_i - Q_{i-1})$ approaches the probability $p(y)\mathrm{d}y$ that y lies in the interval $(y, y+\mathrm{d}y)$. The weighting function in this case can be chosen as

$$w(y) = p(y)/p^*(y) \tag{2}$$

When m tends toward infinity we obtain the "importance sampling technique", which generally permits higher precision and at the same time is more flexible than the stratified sampling.[4]

We now illustrate the general principles in the application of importance sampling to computing the error probability or the false alarm probability of a numerical system. For this we suppose that the output y_n at a certain instant $t = nT$ is a function of actual sample x_n and of the $(n-1)$ previous samples x_i for $1 \leqq i \leqq n-1$. Then writing $\boldsymbol{x} = (x_i \ldots x_n)$, the output is represented by $y_n = g(\boldsymbol{x})$. For example the x_i can be the values in the distance cells of a CFAR signal processor for radar systems.

To compute the false alarm probability P_{fa}, the output y_n of the system is compared with a threshold V. If y_n is greater than V we have a false alarm. Then using the normal Monte Carlo simulation, after N_R processed samples the estimated P_{fa} is

$$P_{\mathrm{fa}} = \frac{1}{N_R} \sum_{i=1}^{NR} U(Z_i - V) \tag{3}$$

where $U(Z_i - V)$ is the unit step function, i.e.

$$U(Z_i - V) = \begin{cases} 0 & \text{if } Z_i < V \\ 1 & \text{if } Z_i \geqq V \end{cases} \tag{4}$$

From the previous results it is clear that the lowest P_{fa}, which can be estimated with N_R observations, is equal to N_R^{-1}. In communication or radar systems, nevertheless, the P_{fa} generally takes low values, lower than 10^{-6}, but often also lower than 10^{-10}–10^{-12}. It is clear that the number of observations required for such a P_{fa} is too high and is impractical for normal Monte Carlo simulation. The principle of the importance sampling technique is to increase the probability of the input $\boldsymbol{x}$ which can give the false alarm. Then, writing $p^*(\mathbf{x})$ for the modified probability of the vector we have a false alarm probability P''_{fa} given by

$$P''_{fa} = \frac{1}{N_R} \sum_{i=1}^{NR} \frac{p(x_i)}{p^*(x_i)} U(Z_i - V) \tag{5}$$

Obviously an important problem is the choice of the distorted probability $p^*(x_i)$. It is clear that this choice depends on the particular system considered, on the required precision and on the allowed processing time.

3. Application of importance sampling techniques to a radar receiver

In the previous section we have illustrated the general principles of importance sampling techniques. Nevertheless the application of these techniques to a real system is not automatic and in general requires a careful choice of the distortion and of the point at which such distortion is applied, together with the weighting function.

In this section we describe two particular importance sampling techniques, which can be applied to many communication and radar systems. These techniques are used to compute the performances of a radar CFAR receiver. Comparison of these results and those obtained from a normal Monte Carlo simulation show that in general these techniques give reasonable accuracy and at the same time a large reduction in the processing time required to compute a given P_{fa}.

The simpler and more intuitive method for applying importance sampling is to distort the input of the system. In particular, considering the evaluation of P_{fa}, we can introduce a higher noise at the input of the system. Then if $p^*(x)$ is the distorted distribution at the input we have a false alarm probability given by equation (5). To characterize the accuracy and the gain in computer time we have applied this method to the system shown in Fig. 2.

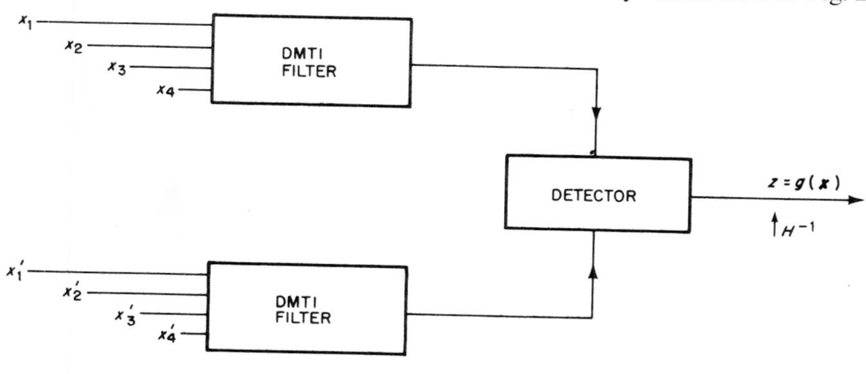

Fig. 2.

In this system, which is a part of a radar CFAR receiver[3,4] we have two identical digital filters of the MTI (moving target indicator) type with time-varying weights and a square-law detector. This structure is considered for its particular characteristics: both linear and non linear devices and time-dependent blocks.

To measure the P_{fa} in this case, we have assumed the threshold is placed after the detector. Curve B in Fig. 3 shows the results obtained for the P_{fa} as a function of the threshold with the importance sampling simulation, while curve A shows those obtained with a normal Monte Carlo simulation. In Fig. 3 the threshold in ordinate is normalized with respect to the r.n.s. σ of the input noise and is labelled V/σ. Clearly the results obtained by using importance sampling are quite accurate. At the same time we have a net reduction of the processing time. This form of importance sampling can be applied without modification to many other communication or radar systems. Nevertheless in some cases this relatively simple approach cannot be used. This happens for example in those systems which contain a block computing the noise level and adapting the output signal to this level. In a CFAR receiver there is a block of this type called the "normalizer" (Fig. 4). In this block the signal in the desired distance cell is normalized with respect to the average noise level computed from

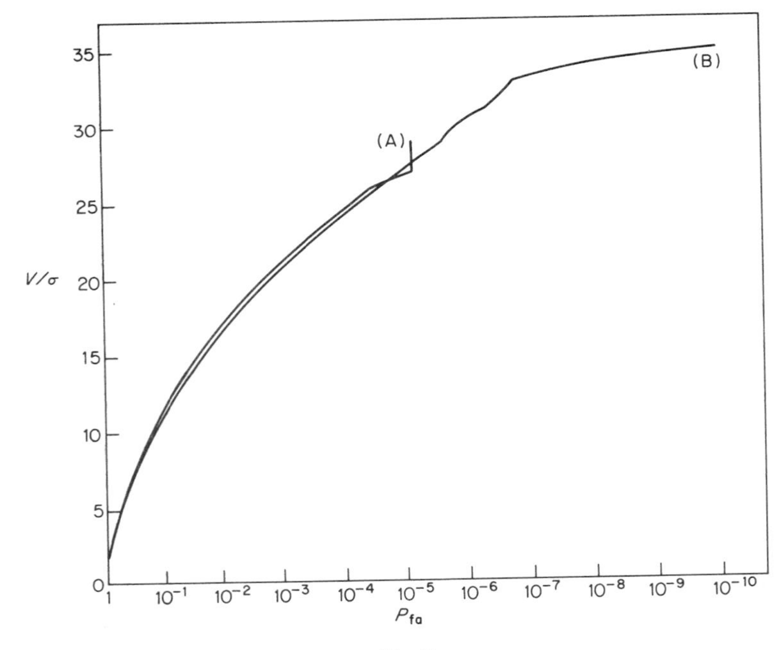

Fig. 3.

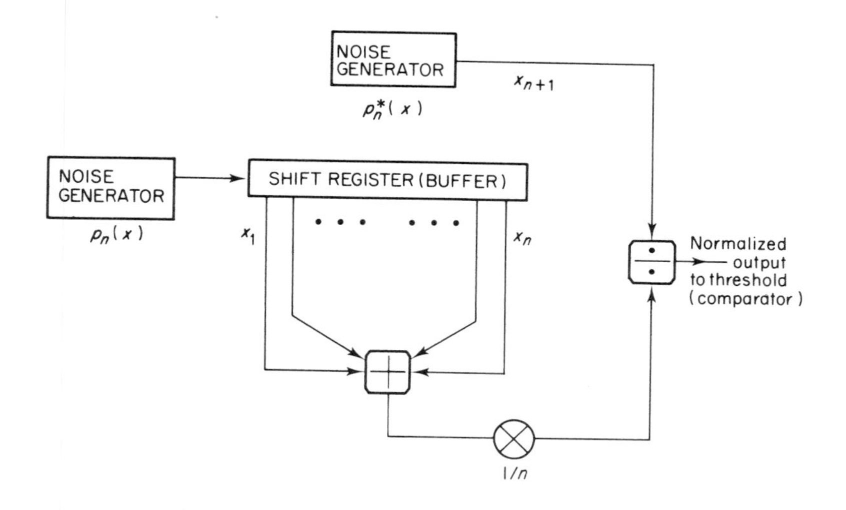

Fig. 4.

some contiguous distance cells.[5] The normalization level can be constructed in many ways. We have considered five different strategies,[5,6] but for brevity in the following we present only the results from one strategy in which the noise level was estimated by averaging the signal in 32 distance cells (16 to the left and 16 to the right of the central distance cell). The output of this block is invariant with respect to variations of the noise. In such a case it is impossible to apply the previous method of importance sampling.

We now describe a second importance sampling technique, which can be applied to these systems. In this method the noise distortion is not introduced at the input of the system, but at some internal point in the receiver. In the case of the system of Fig. 4, for example, only the noise of the central cell is distorted, while the noise of the contiguous cells is the actual noise. After normalization and comparison with the threshold V, the distorted noise is erased from the system and does not enter in the other distance cells of the normalizer. Writing x_j for the distorted noise in the central cell and $p^*(x_j)$ for the distorted probability distribution, we obtain a weighting function

$$w(\boldsymbol{x}) = p(x_j)/p^*(x_j) \tag{6}$$

where $p(x_j)$ is the distribution probability without distortion.

Figure 5 shows some results obtained applying this method to the system in Fig. 4. In particular curve A refers to a normal Monte Carlo simulation, while curves B and C refer to this second importance sampling method

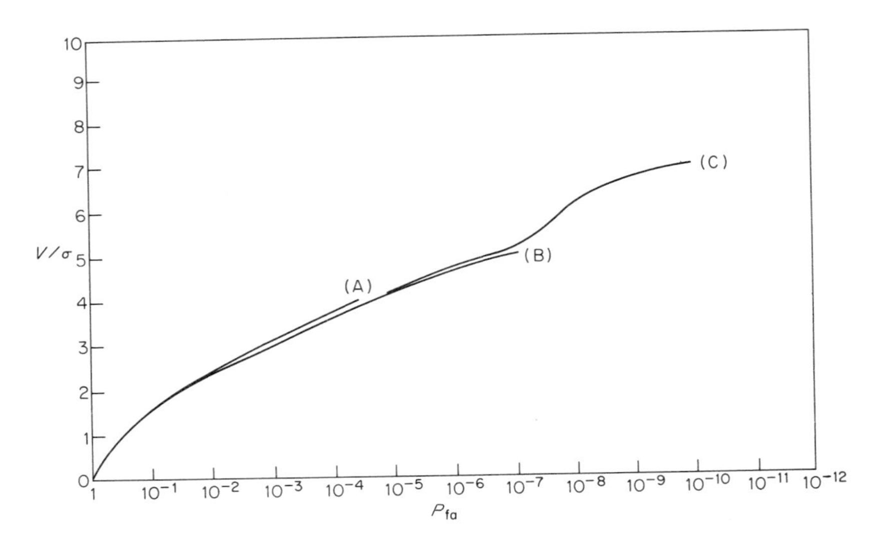

Fig. 5.

with two different distorted probability functions $p^*(x_j)$.† Again in this case, as is clear from the figure, the results obtained using importance sampling are quite accurate. The differences between curve A and curves B and C are greater for P_{fa} less than 10^{-4} where the normal Monte Carlo technique is not reliable (due to an insufficient (10^5) number of processed data). At the same time it is possible with the importance sampling technique, to estimate very low P_{fa} values with the same number of samples as the normal Monte Carlo simulation and the required computer time is reduced by an order of magnitude.[5]

This second technique is also used to evaluate the performance of the digital MTI radar,[4] shown in Fig. 6. Here also the distorted noise was introduced in the central cell of the normalizer and outcomes recognized as false alarms were weighted according to equation (6).

Figure 7 shows the results for P_{fa} for the complete receiver. Again curve A represents the results obtained through a normal Monte Carlo simulation and by using 10^5 samples. As before the curves obtained from the normal Monte Carlo and importance sampling simulation are quite similar, even if the simulation receiver has many hard nonlinearities, resulting from the detector, saturations, thresholds, etc.

To complete the evaluation of the performance of the system we also

† The curves are obtained by processing typically 10^5 samples.

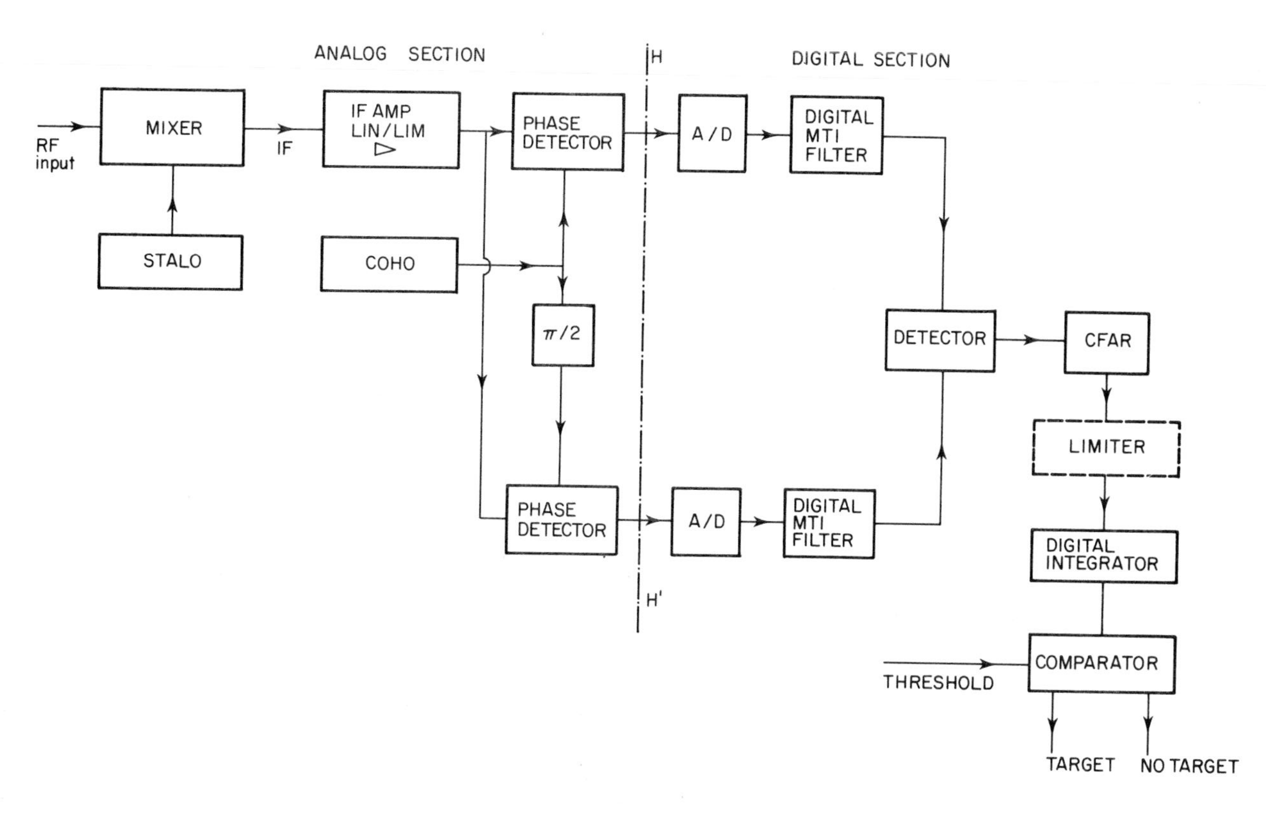

ANALOG SECTION
H
DIGITAL SECTION
MIXER
RF input
IF
IF AMP LIN/LIM
PHASE DETECTOR
A/D
DIGITAL MTI FILTER
STALO
COHO
π/2
DETECTOR
CFAR
LIMITER
PHASE DETECTOR
A/D
DIGITAL MTI FILTER
DIGITAL INTEGRATOR
H'
THRESHOLD
COMPARATOR
TARGET
NO TARGET

Fig. 6.

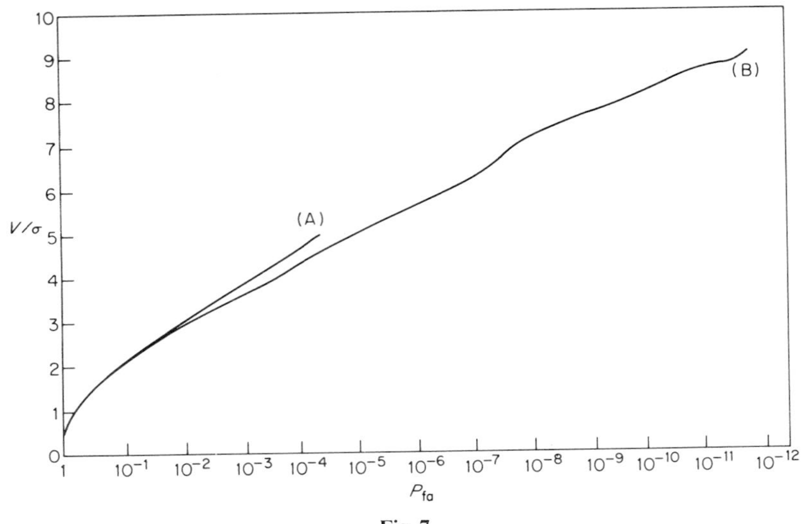

Fig. 7

computed the detection probability P_d, i.e. the probability of detection of the signal in presence of the noise with a given P_{fa}. Because the probabilities of interest are of the order of 0·40 to 0·99, their evaluation can be easily determined by the normal Monte Carlo simulation. From these results it is possible to deduce the signal-to-noise ratio required for a given P_d and from the previous figures also the necessary threshold.

The two importance sampling techniques described in this section are generally applicable to many other communication systems, even if, as for similar techniques discussed in the literature, a careful knowledge of the system and its characteristics is required.

References

1. H. Kahn, "Use of Different Monte Carlo Sampling Techniques", *Symposium on Monte Carlo Methods*, pp. 146–167. New York; Wiley, 1956.
2. B. H. Cantrell and G. V. Trunk, "Importance Sampling", *IEEE Trans. on Aerospace and Electronic Systems*, pp. 878–880 (Nov. 1974).
3. G. V. Hansen, "Importance Sampling in Computer Simulation of Signal Processors", *Comput. and Elect. Eng.*, 1.545.550 (1974).

4. R. Pieroni, "Studio e Simulazione di sistemi radar", Tesi di Laurea in Ingegneria Elettronica, Facoltà di Ingegneria, Università di Firenze, July 1977.
5. Report on "Statistical Evaluation for a Radar Receiver by using Importance Sampling Techniques", Istituto di Elettronica, Facoltà di Ingegneria, contract SMA, Florence, 1978.
6. M. Skolnik, "Radar Handbook". New York; McGraw Hill, 1970.

Digital Chirp Filtering using the Chinese Remainder Theorem

R. C. CREASEY

European Space Research and Technology Centre, Noordwijk, The Netherlands

1. Introduction

Chirp signals, that is signals with a linear frequency variation with time are widely encountered in radar systems. Matched filtering of such signals requires the implementation of chirp filters. Chirp filters can be realized digitally, either by time-domain convolution, or by the use of FFT based convolution. In a time-domain realization Bluestein[1] noted that there is a significant amount of structure in the chirp function which can be harnessed to reduce the amount of processing.

2. Time-domain realization of chirp filters

Sampling an analogue linear frequency-modulated signal with a high time–bandwidth product at the Nyquist rate gives a sampled waveform of

$$\begin{aligned} s(k) &= \exp(-j\pi k^2/N) \qquad && k=0,\ 1,\ 2,\ \ldots\ N-1 \\ &= 0 && \text{elsewhere} \end{aligned}$$

The sampled impulse response of the matched filter to such a waveform, the chirp filter, is then given by

$$\begin{aligned} h(k) &= \exp(j\pi k^2/N) \qquad && k=0,\ 1,\ 2,\ \ldots\ N-1 \\ &= 0 && \text{elsewhere} \end{aligned}$$

A direct transversal filter implemenation of such an impulse response would require N multiplications and N additions per output point. However as noted in ref. 1, many values of $h(k)$ are identical due to the periodicity of $\exp(j\pi k^2/N)$ and thus the number of multiplications can be reduced. Bluestein showed that the number of multiplications required is dependent on the factorization of N, and is equal to the number of distinct

TABLE 1. Number of complex multiplications as a function of N

Factors of N	Number of complex multiplications required	
$N=P$ (prime)	$P-1/2$	
$N=P^t$	$(P^{t+1}-1)/2(p+1)$	t odd
$N=P^t$	$(p^{t+1}-p)/2(p+1)$	t even
$N=2^t$	$(2^{t-2}-1)/3$	t odd
$N=2^t$	$(2^{t-2}-2)/3$	t even
$N=\Pi p_i$	$(\Pi(p_i-1)/2)-1$	

values of a such that

$k^2=a$ Mod N (N odd) has a solution
$k^2=a$ Mod $N/2$ (N even) has solution

Any a which is a solution is said to be a quadratic residue Mod N or Mod $N/2$. Table 1 gives the number of solutions for various factorizations of N (taken from ref. 1). From the table it can be seen that the number of multiplications, per output point, is still proportional to N, although the constant of proportionality varies from $\frac{1}{2}$ for N prime to $\frac{1}{12}$ for N a power 2. It is also to be noted that the number of additions is not reduced. To reduce the number of multiplications and additions further, Bluestein proposed to use a two-dimensional mapping

$$k=k_1+k_2m \qquad k_1,\ k_2=0,\ 1,\ ..\ m-1$$

for values of $N=m^2$. To extend this approach to values of N other than m^2, an alternative mapping is suggested.

3. Chirp filter based on the use of the Chinese Remainder theorem

For those values of k, say k^1 and k^{11}, for which the chirp impulse responses $h(k)=\exp(j\pi k^2/N)$ are equal, i.e. $h(k^1)=h(k^{11})$ then

$$(k^1)^2=(k^{11})^2 \text{ Mod } 2N$$

Thus the evaluation of the indices can be made Modulo $2N$.

Considering the case where $N=n_1n_2$, with n_1 and n_2 being mutually prime and odd, then by the Chinese Remainder theorem there is a unique mapping of the integer k Mod $2N$ onto the integers k_1 Mod n_1, k_2 Mod n_2 and k_3 Mod 2 given by

$$k=ak_1+bk_2+ck_3 \text{ Mod } 2N$$

where

$$a=\frac{2N}{n_1}\left(\frac{2N}{n_1}\text{ Mod } n_1\right)^{-1} \qquad b=\frac{2N}{n_2}\left(\frac{2N}{n_2}\text{ Mod } n_2\right)^{-1} \qquad c=N$$

Using this mapping gives

$$\begin{aligned}k^2&=(ak_1+bk_2+ck_3)^2 \text{ Mod } 2N\\ &=a^2k_1^2+b^2k_2^2+c^2k_3^2+2abk_1k_2+2ack_1k_3+2bck_2k_3 \text{ Mod } 2N\end{aligned}$$

but

$$2\ ab=2ac=2bc=0 \text{ Mod } 2N$$

$$a^2=a \text{ Mod } 2N \qquad b^2=b \text{ Mod } 2N \qquad c^2=c \text{ Mod } 2N$$

thus

$$k^2=ak_1^2+bk_2^2+ck_3^2$$

and hence

$$\exp(\mathrm{j}\pi k^2/N)=\exp(\mathrm{j}\pi ak_1^2/N)\ \exp(\mathrm{j}\pi bk_2^2/N)\ \exp(\mathrm{j}\pi ck_3^2/N)$$

However $c=N$ and k_3 only takes on the values 0 and 1, thus $\exp(\mathrm{j}\pi k_3^2)$ equals $+1$, $(k_3=0)$, and -1, $(k_3=1)$, that is $\exp(\mathrm{j}\pi k_3^2)=(-1)^{k_3}$. Furthermore since we are only interested in values of exp $(\mathrm{j}\pi k^2/N)$ for $0\leqq k\leqq N-1$, then over this range k Mod $2=((ak_1+bk_2)$ Mod $N)$ Mod 2. So $(-1)^{k_3}=(-1)^k=(-1)^p$ where $p=(ak_1+bk_2)$ Mod N. Thus the explicit multiplication by $\exp(\mathrm{j}\pi k_3^2)$ can be replaced by sign inversions, enabling the one-dimensional convolution.

$$y(n)=\sum_{k=0}^{N-1} x(n-k)\ \exp(\mathrm{j}\pi k^2/N)$$

to be written as a two-dimensional convolution with a separable kernal, i.e.

$$y(n)=\sum_{k_1=0}^{n_1-1}\exp(\mathrm{j}\pi ak_1^2/N)\sum_{k_2=0}^{n_2-1}(-1)^p x(n-(ak_1+bk_2)\text{Mod } N)\exp(\mathrm{j}\pi bk_2^2/N)$$

This two-dimensional convolution can then be performed by first calculating the n_2 inner sums, (over k_2), and then the final sum, (over k_1). Consider for example, the case where $N=15$, i.e. $n_1=5_1$, $n_2=3$, $a=5$, $b=10$. Then using the above mapping and setting $W=\exp(\pi\mathrm{j}/N)$, successive outputs of the convolution are given by,

$$\begin{aligned}y(n))=\ &1[x(n)W^0-x(n-5)W^{10}+x(n-10)W^{10}]\\ &-w^6[x(n-1)W^{10}-x(n-6)W^0+x(n-11)W^{10}]\\ &+W^{24}[x(n-2)W^{10}-x(n-7)W^{10}+x(n-12)W^0]\\ &-W^{24}[x(n-3)W^0-x(n-8)W^{10}+x(n-13)W^{10}]\\ &+W^6[x(n-4)W^{10}-x(n-9)W^0+x(n-14)W^{10}]\end{aligned} \quad \Big\downarrow k_1$$

$$
\begin{aligned}
y(n+1) = 1[&x(n+1)W^0 - x(n-4)W^{10} + x(n-9)W^{10}] \\
&- W^6[x(n)W^{10} - x(n-5)W^0 + x(n-10)W^{10}] \\
&+ W^{24}[x(n-1)W^{10} - x(n-6)W^{10} + x(n-11)W^0] \quad k_1 \downarrow \\
&- W^{24}[x(n-2)W^0 - x(n-7)W^{10} + x(n-12)W^{10}] \\
&+ W^6[x(n-3)W^{10} - x(n-8)W^0 + x(n-13)W^{10}]
\end{aligned}
$$

From this it can be seen that for a given $y(n)$ each inner sum over k_2 is formed by the dot product of a different set of n_2 samples of $x(k)$, spaced by n_1, and a cyclically shifted version of the impulse response $\exp(j\pi bk_2^2/N)$. A given sample set, (e.g. $x(n)$, $x(n-5)$, $x(n-10)$ above), contributes to n_1 successive outputs ($y(n)$ to $y(n+n_1)$, and its contribution consists of the successive outputs of the cyclic convolution of $\exp(j\pi bk_2^2/N)$ with that sample set. Furthermore if $n_1 > n_2$ all of the outputs of the cyclic convolution are used and $n_1 - n_2$ are used more than once. This implies that instead of computing the n_1 inner sums directly for each $y(n)$, the length n_2 cyclic convolution of $x(n)$, $x(n-n_1)$, etc., with $\exp(j\pi bk_2^2/N)$ can be computed and the n_2 outputs saved. At any one time n_1 sets of these intermediate results (each of n_2 values) must be saved. The final output is then formed by taking one result from each of the n_1 sets, multiplying each result by the appropriate value of $\exp(j\pi ak_1^2/N)$ and forming the outer sum. Thus for each output, the n_2 outputs of a length n_2 convolution must be computed, which requires $\sim n_2^2$ complex additions and up to n_2^2 complex multiplications. Taking into account the n_1 additions and multiplications required for the outer sum leads to a total number of operations proportional to $n_1 + n_2^2$, compared to $n_1 n_2$ per output point of the direct method. It can also be shown that the computational requirements for this mapping are minimized when $n_2^2 \cong n_1$.

Further improvements to this basic method can be found, by improving the efficiency of computing the length n_2 inner cyclic convolution. For example Table 1 can be used to determine the minimum number of multiplications to be used. Alternatively rather than compute the inner cyclic convolution directly, use can be made of transform techniques, such as the FFT or number theoretic transforms. In such cases, the number of operations can be reduced for favourable values of n_2, i.e. n_2 highly composite. However when n_2 is composite then N can be written as $N = n_1 n_2 n_3 \ldots n_j$, and hence the two-dimensional index mapping given previously can be extended to higher, j-dimensional mapping, i.e.

$$k = a_1k_1 + a_2k_2 + a_3k_3 + \ldots + a_jk_j \text{ Mod } 2N$$

and

$$k^2 = a_1k_1^2 + a_2k_2^2 + a_3k_3^2 + \ \ldots \ a_jk_j^2 \text{ Mod } 2N$$

with the k_i being evaluated Mod n_i and the n_i being mutually prime. Under this mapping the original one dimensional convolution can be rewritten as a multidimensional convolution:

$$y(n) = \sum_{k_1=0}^{n_1-1} \exp(j\pi a_1 k_1^2/N) \sum_{k_2=0}^{n_2-1} \exp(j\pi a_2 k_2^2/N) \ldots$$

$$\sum_{k_j=0}^{n_j-1} x(n - a_1 k_1 \ldots a_j k_j) \exp(j\pi a_j k_j^2/N)$$

In this case again computation of the inner (k_jth dimension) cyclic convolution and saving intermediate results leads to a number of operations proportional to $n_1 + n_2 + n_3 + \ldots n_j^2$.

Returning to the original formulation of the problem in which n_1 and n_2 were not only mutually prime but also both odd, led initially to a three-dimensional mapping, Mod n_1, Mod n_2 and Mod 2, which was reduced, by essentially changing the sign of alternate data samples $x(n)$, to a two-dimensional mapping Mod n_1, Mod n_2. In the case where, say n_2 is even, then 2 and n_2 are no longer mutually prime and the mapping is made immediately into two dimensions, i.e.:

$$k = ak_1 + bk_2 \text{ Mod } 2N$$

where k_1 is evaluated Mod n_1 and k_2 is evaluated Mod $2n_2$. In this case the sign change of the data samples is not required. Also note that since $2n_2$ has 4 as a factor, then $(k_2)^2 = (k_2 + n_2)^2$ Mod $2n_2$. Finally Fig. 1 indicates an implementation of a two-dimensional chirp filter based on the Chinese Remainder theorem.

4. Chirp filter for $N = 2^p$

The procedure in the previous sections applies when N is highly composite, and the factors of N are mutually prime. When $N = 2^p$, the factors are not mutually prime and the Chinese Remainder theorem mapping cannot be used. However it is possible to apply a decimation procedure as is done for the FFT as follows:

$$y(n) = \sum_{k=0}^{N-1} x(n-k) \exp(j\pi k^2/N) \qquad N = 2^p$$

$$= \sum_{k=0}^{N/2-1} x(n-k) \exp(j\pi k^2/N) + \sum_{k=N/2}^{N-1} x(n-k) \exp(j\pi k^2/N)$$

$$= A(n) + B(n)$$

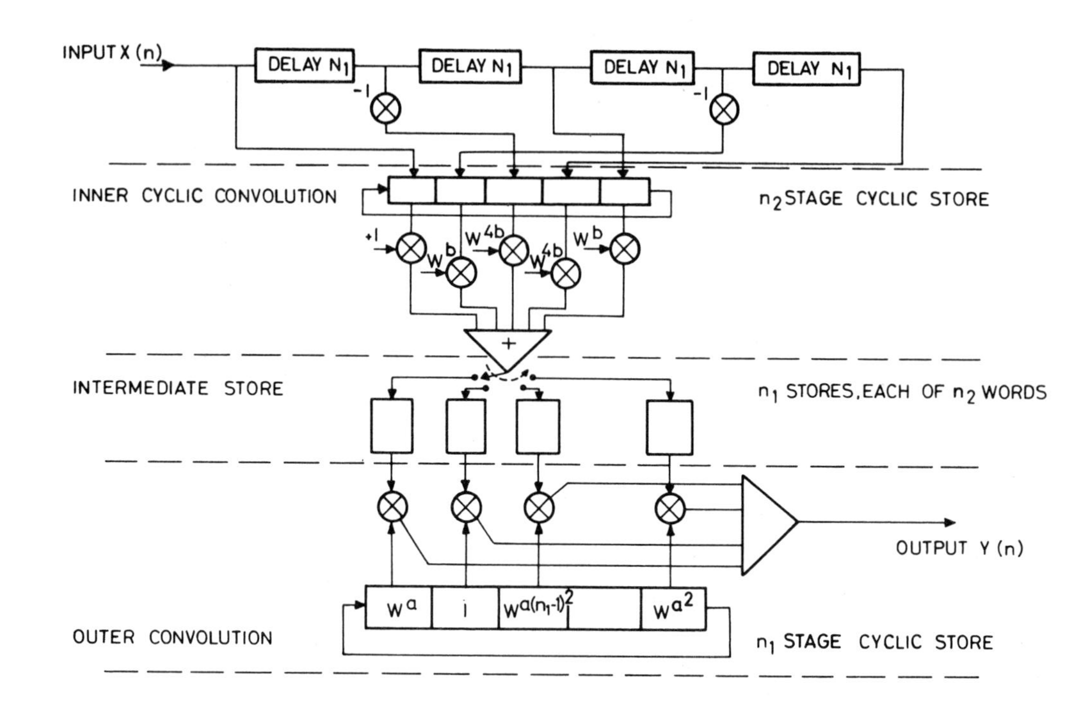

Fig. 1. Chirp filter for $N = n_1 n_2$.

Setting $k=l+N/2$ gives

$$B(n)=\exp(j\pi 2^{p-2})\sum_{l=0}^{N/2-1} x(n-N/2-l)\exp(j\pi l)\exp(j\pi l^2/N)$$

Since $\exp(j\pi l)$ takes on the value $+1$, l even, and -1 for l odd, then

$$B(n)=\exp(j\pi 2^{p-2})\left[\sum_{l=0,2,4}^{N/2-2} x(n-N/2-l)\exp(j\pi l^2/N) - \sum_{l=1,3}^{N/2-1} x(n-N/2-l)\exp(j\pi l^2/N)\right]$$

$$=\exp(j\pi 2^{p-2})[B\text{ even }(n)-B\text{ odd }(n)]$$

but B even $(n)=A$ even $(n-N/2)$, B odd $(n)=A$ odd $(n-N/2)$. Thus, for $p\geqq 3$,

$$y(n)=A\text{ even }(n)+A\text{ even }(n-N/2)+A\text{ odd }(n)-A\text{ odd }(n-N/2)$$

If the values of A even $(n-N/2)$ and A odd $(n-N/2)$ are saved from the computation of $y(n-N/2)$, then only A even (n) and A odd (n) need be computed, which requires a total of $N/2$ operations per output point compared with N by the direct method.

As with the FFT further stages of decimation are possible, for example the second stage of decimation is:

$$A(n)=\sum_{k=0}^{N/2-1} x(n-k)\exp(j\pi k^2/N)$$

$$=\sum_{k=0}^{N/4-1} x(n-k)\exp(j\pi k^2/N)+\sum_{N/4}^{N/2-1} x(n-k)\exp(j\pi k^2/N)$$

$$=C(n)+D(n)$$

setting $k=l+N/2$ and substituting gives

$$\exp(j\pi k^2/N)=\exp(j\pi N/16)\exp(l/2)(\exp\, j\pi l^2/N)$$

For values of $p\geqq 5$, $\exp(j\pi N/16)=1$. Also $\exp(j\pi l/2)$ takes on four distinct values only, namely $+1$, -1, j, $-j$. Thus $D(n)$ can be written,

$$D(n)=d_0(n)-d_2(n)+j[d_1(n)-d_3(n)]$$

where

$$d_i(n)=\sum_{l=i,i+4...}^{N/4-i} x(n-N/4-l)\exp(j\pi l^2/N) \qquad (i=0, 1, 2, 3).$$

Since $C(n)$ can also be written

$$C(n)=C_0(n)+C_2(n)+C_1(n)+C_3(n)$$

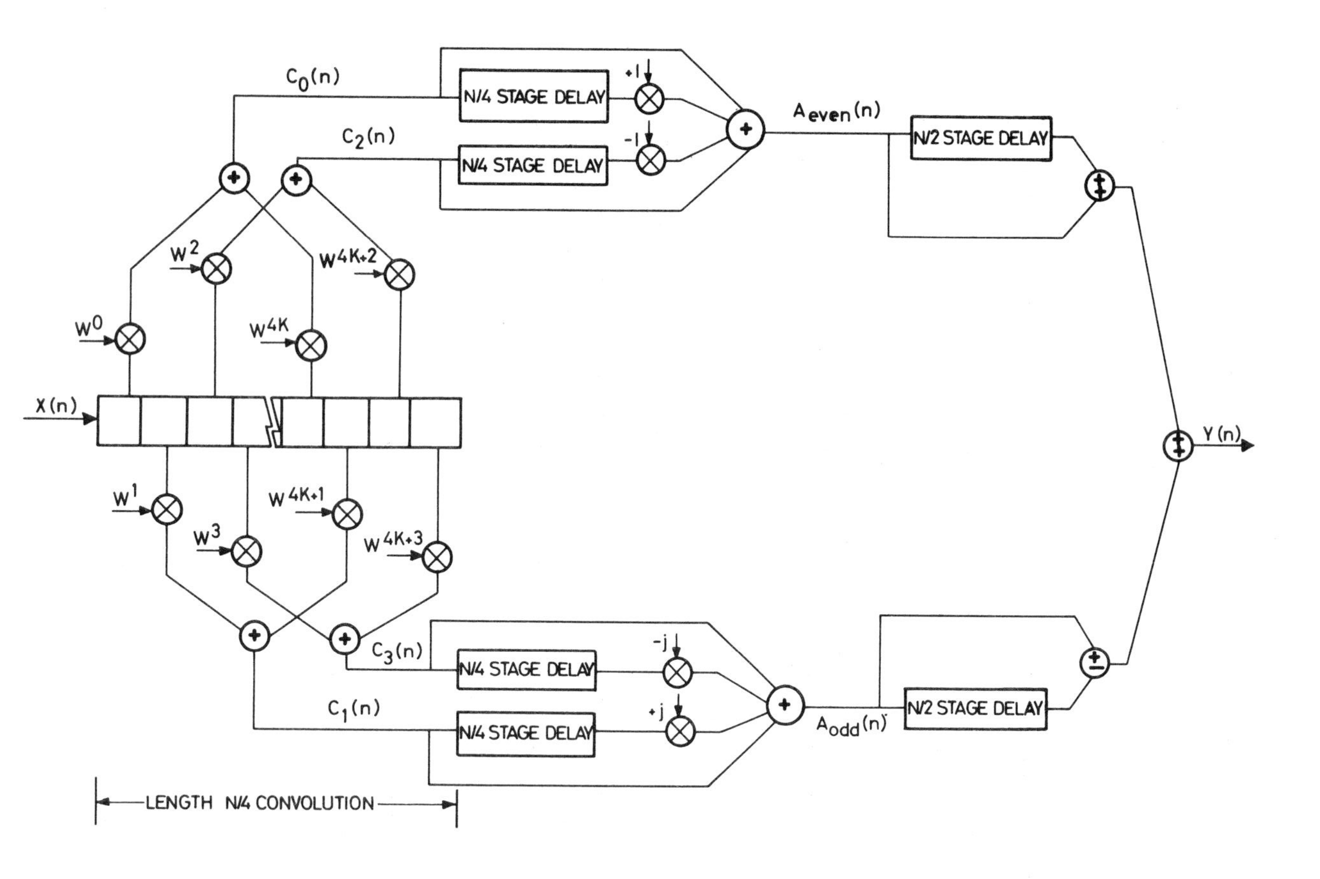

Fig. 2. Chirp filter structure $N = 2^p$, 2-stage decimation.

where

$$c_i(n) = \sum_{l=i,i+4,}^{N/4-i} x(n-l)\exp(j\pi l^2/N)$$

then $d_i(n) = C_i(n - N/4)$ and

$$A \text{ even } (n) = C_0(n) + C_2(n) + C_0 n - N/4) - C_2(n - N/4)$$

$$A \text{ odd } (n) = C_1(n) + C_3(n) + j[C_1(n - N/4) - C_3(n - N/4)]$$

Again if the values of $C_i(n - N/4)$ are saved from the computation of $A(n - N/4)$ then only the four $C_i(n)$s need be computed. Since each $C_i(n)$ is of length $N/16$, the number of operations per output point is approximately $4(N/16) = N/4$.

Similarly further stages of decimation lead to a number of operations proportional to $\log_2 2^p = p$ per output point.

An implementation of a two-stage decimated chirp filter is sketched in Fig. 2. As is the case for Bruens[2] FFT filters, each stage of decimation adds N extra memory cells.

5. Conclusions

The procedures for the computation of chirp filters presented here permit efficient direct, that is without the use of the FFT, realizations of the sampled chirp filter to be achieved. Since the forward and inverse transforms of an FFT are not required, the proposed procedures will in many cases be more efficient than an FFT based realization.

References

1. L. Bluestein, "A linear Filtering Approach is the Computation of the Discrete Fourier Transform," *IEEE Trans. on Audio and Acoustics.* (Dec. 1970).
2. G. Bruen, "Z Transform Filters and DFT's." *IEEE Trans. on Acoustics Speech and Signal Processing.* (Feb. 1978).

Performance Evaluation of an Interrogation–Reply Scheduling Technique for a Discrete Address Beacon System

GIACOMO BUCCI and DARIO MAIO

CIOC-CNR and University of Bologna, Italy

1. Introduction

In today's air traffic control systems, surveillance is provided mainly by a co-operative interrogator/transponder system, secondary surveillance radar (SSR), or, particularly in the United States, the Air Traffic Control Radar Beacon System (ATCRBS). Communication between the pilot and controller is by UHF voice radio.

In recent years, a discretely-addressed interrogation technique, called DABS (discrete address beacon system), has been developed at Lincoln Laboratory. The fundamental difference between DABS and ATCRBS is the manner of addressing aircraft, or selecting which aircraft will respond to an interrogation. In DABS, each aircraft is assigned a unique address code; the selection of which aircraft is to respond to an interrogation is accomplished by including the aircraft's address code in the interrogation. The two major advantages accruing from the use of discrete address for surveillance are as follows.[1,2]

Firstly, an interrogator is now able to limit interrogation to only those targets for which it has surveillance responsibility, rather than continuously interrogate all targets within line-of-sight. This prevents surveillance system saturation caused by all transponders responding to all interrogators within line-of-sight. Secondly, appropriate timing of interrogations ensures that the responses from aircraft do not overlap, eliminating the mutual interference which results from the overlapping of replies from closely spaced aircraft (termed "synchronous garble"). In addition to improved surveillance capability, use of the discrete address in interrogation and replies permits the inclusion of messages to or from a particular aircraft, thereby providing the basis for a ground–air and air–ground digital data link.

However, DABS surveillance and communication involve greater hardware and software complexity in channel management. Namely, DABS sensors must operate with two classes of transponders, i.e. SSR and DABS transponders which, therefore, share the same RF channel for uplink and downlink messages. As a result, the program-controlling RF channel, must perform both SSR and DABS activity, alternating them in the time domain. While the conventional radar/beacon systems do not require special control procedures, execution of DABS surveillance activity introduces great complexity for the following reasons:

(1) DABS interrogation addresses single aircraft discretely, therefore DABS sensors must be able to predict when the aircraft will be within the antenna beam.

(2) Channel time must be allocated to each individual DABS interrogation and reply, therefore a prediction of aircraft range is required.

(3) DABS surveillance procedures often require more than one interrogation per aircraft and, if a given interrogation fails to produce a usable reply, the sensor must repeat the attempt as long as the aircraft remains in the beam.

This paper presents some results of a study carried out as part of an ATC project sponsored by Italian CNR; the aim of the project is to develop an improved air traffic control system. The main objectives of the subproject to which this work belongs are listed below.

(a) To implement the ATC system on mini or micro computers.

(b) To give an account of the maximum supported target load, which is expected to be quite low in comparison with that supported by the surveillance system described in refs. 1 and 2.

(c) To achieve the greatest reliability of the surveillance and communication functions performed by ATC.

The first part of this paper briefly describes the structure of a real-time system proposed for channel management. The second part deals with the major surveillance functions and develops a number of design criteria to reduce the associated computation effort. Finally, an evaluation of interrogation–reply scheduling techniques is presented.

2. Channel management

This section describes both hardware and software structures of the computer system performing channel management.

The functional architecture of the sensor is essentially the same as that adopted at Lincoln Laboratory. A block diagram showing information flow is given in Fig. 1. System components and associated functions shown in Fig. 1 are as follows.

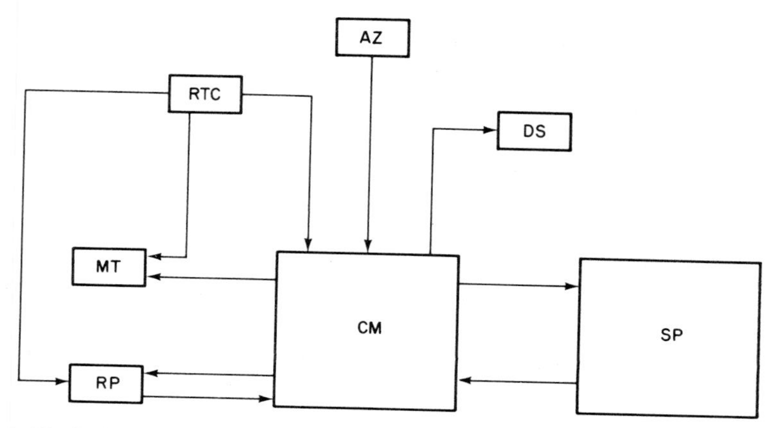

Fig. 1. Block diagram of DABS showing information flow. (See text for description of components and functions.)

CM: is the computer that controls the RF channel, schedules transactions for targets within the antenna beam, handles information and regulates all system interactions.

SP: is a computer that maintains and updates the surveillance file which, for each DABS target under control, keeps track of the predicted position and the pending uplink messages and downlink message requests. SP transfers to CM data blocks containing complete specification of the required transactions for targets about to enter the antenna beam, and receives information from CM on targets leaving the beam, this being used to update the surveillance file. Other functions performed by SP are not discussed in this paper.

MT: is the modulator-transmitter. It accepts commands from CM and generates the requisite RF interrogation signals. A command contains information for building the message to be transmitted to a target and the instant of time at which transmission must be started.

RP: is the reply processor. It accepts commands from CM in order to receive and process the targets' replies that have been elicited by previous calls transmitted by MT. Processed replies are then transferred to CM. CM has complete control over MT and RP. CM communicates with these units by means of a stream of interrogation and reply control commands and receives from RP a stream of DABS reply data blocks.

AZ: is the antenna azimuth register. It gives, at any instant of time, the azimuth of the rotating antenna. AZ is directly read by CM.

RTC: is the system real-time clock. It can be directly read by MT, RP and CM.

DS: is a video-display to communicate messages and error conditions to the operator.

The system software is structured as a set of cooperating sequential processes.[3] In our conceptualization, a process is the largest set of sequentially executed actions which could be performed by an independent processor.

There are two kinds of processes in the system: external and internal:

External processes are those which, besides interacting with other processes, must also synchronize with the activities of external devices. External processes are called drivers.

Internal processes are those which interact with the other system processes, but never synchronize with an external device. There is a driver for each independent external device and an internal process, called channel control (CC), which coordinates all the others and performs computation functions. Process coordination is obtained through semaphores,[3] queue semaphores,[4] synchronization primitives and other kernel primitives. A more detailed description of process structure and process interaction can be found in ref. 5.

3. Software functions for channel management

This section describes the major channel management functions required by DABS surveillance.

The DABS sensor maintains a regularly updated list of DABS targets within the antenna beam and utilizes the DABS time to make repeated passes through this list (called active target list), scheduling discretely-addressed DABS interrogations and replies on a nonconflicting basis. To this end the following functions must be performed. The first of them is implemented on SP, whereas the remaining are executed on CM as a part of the CC process.

Transaction preparation

Before targets enter the antenna beam, transaction preparation produces the associated new target transaction blocks. The basis of this activity is information stored in the surveillance file which contains the so-called "active message list" of pending uplink message and downlink message requests. The surveillance file is azimuth ordered and transaction preparation is given a new azimuth limit (Θ_{new}) each time it is activated. Transaction preparation retrieves from the file those targets whose predicted azimuths fall between the new and previous azimuth limit.

Target list update
Transaction scheduling for a DABS interrogation period is based on information contained in the active target list. CC updates this list before entering a DABS period with new target transaction blocks obtained from SP. To have these data, CC communicates the related Θ_{new} to SP in due time (i.e., before the beginning of the DABS period preceding that for which new targets data are required.

Another responsibility of this function is to delete the block completed targets and those which have fallen behind the beam, thereby forming a released target list to be sent to SP for surveillance file updating.

Roll-call scheduling
This is the most important function performed by channel management, since it produces a schedule of interrogation and reply cycles. Cycles are computed based on the available time, pending transactions and delay time between interrogations and related replies.

Transaction update
Active target list is updated by CC on the basis of replies to previous interrogations. Transaction update is usually performed after a schedule computation.

The previous discussion is a very general presentation of the main functions performed for channel management. The way these functions should be implemented and the associated data structures have been studied in detail in ref. 6. The remaining part of this paper compares expected performance of the solution of ref. 6 with that of ref. 2.

4. Design considerations

We list the main characteristics of the DABS system described in ref. 2:

(a) The ability to provide discrete-address surveillance of more than 2000 aircraft.

(b) The decreasing range ordering of the active target list which allows the use of a high packing efficiency algorithm (Full-Ring[7]). However, maintaining the range-ordered structure of the active target list determines a relatively high computation overhead in list updating, both for merging the new targets and forming the released target list.

(c) Referring to the scheduling algorithm, the absence of gap between successive interrogations, with the exception of that due to range guard, and the minimum spacing imposed to assure resuppression of ATCRBS transponders with each DABS interrogation.

From the system point of view, packing efficiency, computation effort and core requirement are important measures of surveillance performance. For a set of scheduled targets, "packing efficiency" is defined as the ratio of the sum of scheduled message lengths (in units of time) to the channel time from the beginning of the first interrogation to the end of the last reply. By choosing to implement the ATC system on minicomputers, computation time becomes the most important parameter.

Given the overall project objectives, the three major design choices were the following:

(1) To design and implement channel management functions in such a way that computation time is reduced to the minimum.
(2) To adopt a rotating fixed-beam antenna.
(3) To use a transmitter with a low duty cycle (about $\leqq 10\%$), a feature that forces a large gap between interrogations.

As a consequence of both previous choices and the low target peak load predicted for the surveillance system, the basic design criteria were the following. Firstly, to reduce the computation effort associated with list updating we had to renounce range-ordering of the active target list. As a result, the active target list is azimuth-ordered, thus no operation is needed when new targets are entered in the active target list from the surveillance file. Secondly, suppressing range-order of the target list imposed the need to design a new scheduling algorithm. For this purpose, a very simple scheduling algorithm, called Fifo, has been proposed.[6] This schedules the targets in range random order.

Finally, further considerations of Full-Ring packing efficiency led to the following conclusions.

(1) The best application of the Full-Ring algorithm is for "agibile" beam sensor.[7] Note that for a rotating beam antenna the scheduler is restricted to scheduling a target in the time interval that the target is covered by the beam, so that only a limited set of targets is to be scheduled in a defined interval of time. Full-Ring packing efficiency degrades with low target numbers.
(2) The performance also degrades with unequal uplink and downlink message length or when a gap is forced between interrogations, as these would increase the time loss on the channel between successive targets. The effect of message length distribution on packing efficiency is described in ref. 7; here the Close-Fit algorithm is proposed to schedule unequal-length messages while maintaining high packing efficiency.

However, of the many algorithms considered in ref. 7 for interrogation–reply scheduling, Full-Ring has been selected in the solution proposed in ref. 2 for a sensor that utilizes a rotating fixed-beam antenna and for the

TABLE 1.

	Short	Long
Interrogation	18·5 μs	32·5 μs
Reply	64·0 μs	120·0 μs

signal durations shown in Table 1. This is because Full-Ring has the following desirable features:[2]

(a) It has flexibility to accomodate several distinct transaction types used in DABS.
(b) It produces efficient schedules, regardless of range distribution of the targets.
(c) Computation is accomplished in a single pass through the active target list.
(d) The targets are added to the schedule in the same order as they are to be interrogated.

5. Evaluation

Fifo and Full-Ring have been evaluated by exercising a large number of times and building statistics for packing efficiency with different duty values.[6] Packing efficiency versus the number of targets scheduled is shown in Figs. 2 and 3 for Full-Ring and Fifo, respectively. Numerical values of signal durations have been assumed with equal occurrence probability to those of Table 1. Range guard was assumed to be 40 μs. Target ranges were randomly generated between 2 and 150 km with uniform distribution.

As the figures show, the difference between Fifo and Full-Ring packing efficiency is small when we consider a number of targets fewer than twelve and a transmitter duty cycle less than or equal to 10%. It may therefore be noted that Full-Ring is always, from the packing efficiency viewpoint, the better way to pack the transactions into the RF channel, owing to the fact that the active target list is range-ordered even if its packing efficiency markedly drops from the 95% obtained with the agibile beam antenna, equal-length messages and absence of a required gap between two successive interrogations. Moreover, it is observed that, given the previously listed parameter values, Full-Ring reaches its maximum packing efficiency for a given number of targets on the list when the duty cycle is approximately 31·25%. This is, in fact, the minimum duty cycle value required to satisfy the worst possible case which occurs when two consecutive scheduled targets have an almost equal range delay.[6] Note that a lower duty

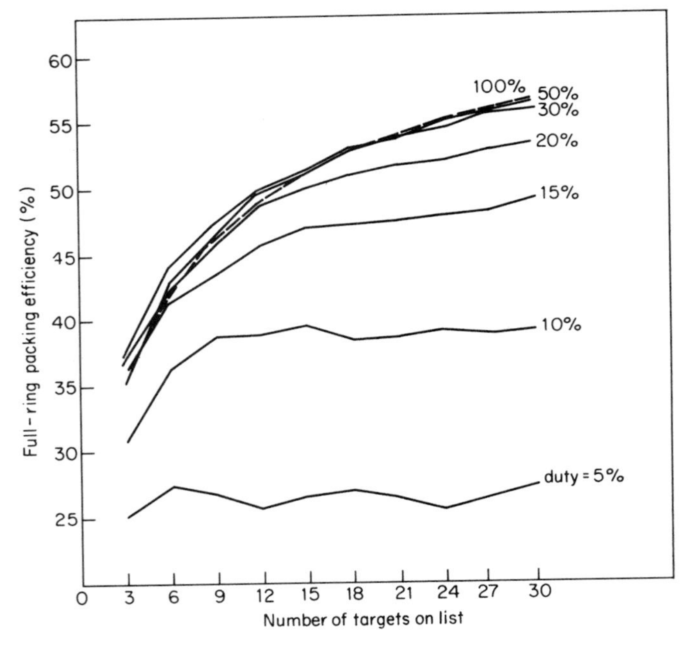

Fig. 2. Packing efficiency versus the number of targets for Full-Ring.

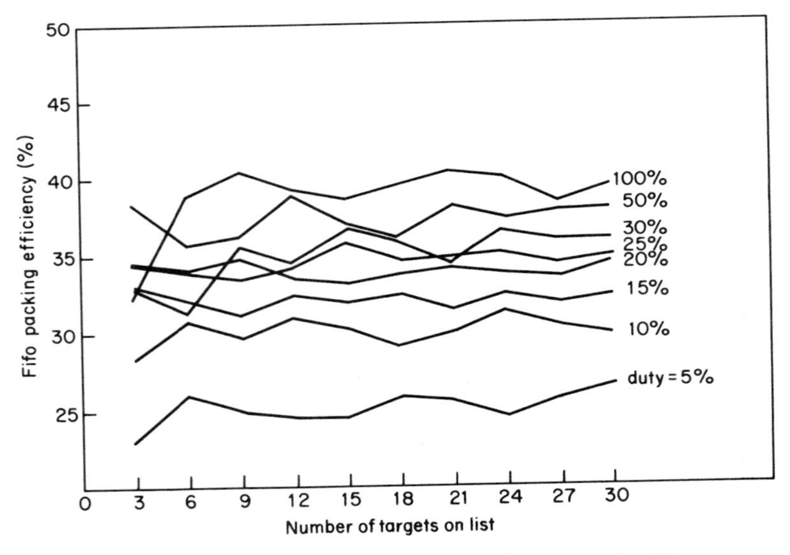

Fig. 3. Packing efficiency versus the number of targets for Fifo.

cycle value would reduce packing efficiency, whereas a higher value cannot increase it.

Let us now consider the minimum spacing (50 μs) imposed to assure the resuppression of ATCBRS transponders. The technique utilized by Full-Ring to pack the transactions into the RF channel assures, on the basis of the values assumed for the reply duration and the range guard, a gap always greater than 50 μs between the leading edges of two successive interrogations. Moreover the technique utilized by Fifo guarantees this only when the duty cycle is less than 37%. In other words, when the signal durations of Table 1 are given, the Fifo plots relative to a duty value greater than 37% have no meaning for a 50 μs required gap.

The loss of packing efficiency of the Fifo algorithm compared to the Full-Ring is the measure of the price to be paid when renouncing a range-ordered list. An advantage though is that the elimination of target list updating represents a saving in computation time which, although small owing to the low target load, may be utilized to guarantee production of an interrogation–reply cycle before the beginning of each discrete interrogation period. In addition the particular list structure employed (circular buffer, always azimuth-ordered) makes formation of the released list extremely simple and fast. In fact, aircraft abandon the list only when, due to antenna rotation, they fall behind the beam. Since the list may be thought of as rotating with the antenna, it is may be seen how formation of the released list is reduced to the updating of only two pointers.

However, new problems relating to transaction updating arise precisely because of the type of data structure utilized.[6] These problems involve a certain computation time overhead; this is spent at the end of the period and is, therefore, less critical. It is further supported by the hypothesis of executing a single schedule for each discrete address interrogation period. Further change with respect to the DABS channel management functions has been achieved by suppressing the allocation calculation task, thus reducing computation time in roll-call scheduling before the beginning of each discrete interrogation period. A suitable data structure permits account to be taken of the priority rules to select those transactions which must be served in the available time on the channel.

We now discuss further simulation results. In order to verify that the Fifo scheduling algorithm guarantees surveillance of the predicted traffic load even under peak conditions, simulation experiments have been carried out to investigate how many normal surveillance transactions can be allocated on the RF channel in a DABS period. A normal surveillance transaction is one in which both interrogation and reply are short. Figure 4 shows the number of normal surveillance transactions per millisecond (NT/ms) as a function of the duration of the DABS period (T_{DABS}) for two duty values of

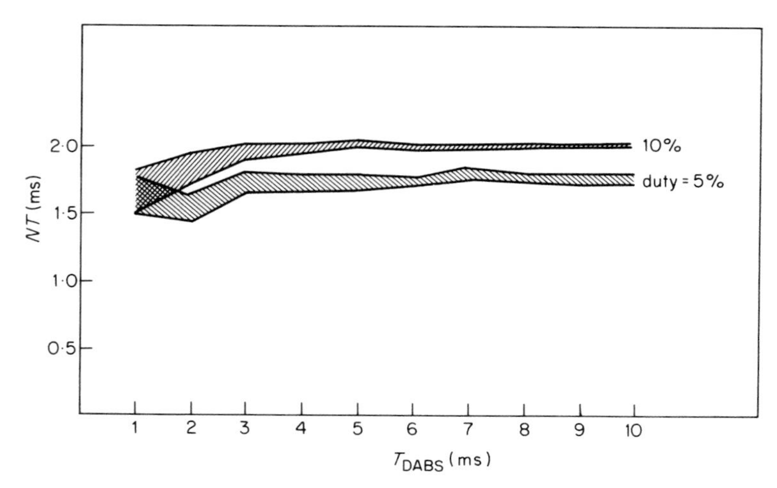

Fig. 4. Number of normal surveillance transactions as a function of the DABS period.

interest and for a range guard $G=40$ μs. Aircraft distance has been assumed to be uniformly distributed in the interval 2–150 km. Figure 4 has two shaded regions corresponding to a 99% confidence interval. Assuming for instance, $T_{DABS}=4$ ms, duty = 5% and taking the lowest value (for a consecutive quantification) of NT/ms in the confidence interval, the number of normal surveillance transactions per millesecond is more than 1·5 and, therefore, the number of these transactions in the DABS period is more than 6. In other words, the maximum number of transactions that can be allocated on the RF channel in a DABS period is $NT_{max}=6$.

To represent the capabilities of the Fifo algorithm completely, the previous figure must be related to the angular spacing in which those transactions are executed. Under peak traffic conditions, a sensor must be able to perform the normal surveillance transaction for each aircraft entering the antenna beam, with no need for other transactions with the same aircraft during the time it remains in the beam. Figure 5 shows two successive positions of the leading edge of the antenna beam relative to the beginning of DABS periods D_1 and D_2. In Fig. 5, S_1, S_2 and S_3 stand for SSR periods. During D_2, the new targets that are to be interrogated are those which entered the beam during the time associated to D_1 and S_2. Under the assumption that DABS and SSR periods have fixed durations, T_{DABS} and T_{SSR} respectively, the amplitude of the angle covering entering aircraft is constant and given by

$$\alpha = 2\pi(T_{DABS} + T_{SSR})/T_{SCAN}$$

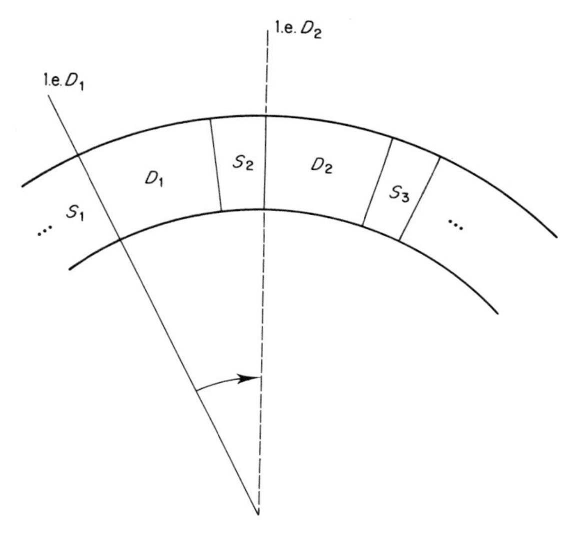

Fig. 5. Two successive positions of the leading edge of the antenna beam. D_1 and D_2 are DABS periods; S_1, S_2, S_3 are SSR periods.

where T_{SCAN} is the time for a complete rotation. Now, assuming again $T_{DABS}=4$ ms, $T_{SSR}=2$ ms and $T_{SCAN}=4S$, we have $\alpha=0{\cdot}5^{\circ}$. Therefore, with the previous assumptions, the Fifo algorithm, can accommodate a maximum of six normal surveillance transactions in half a degree.

Acknowledgment

This work has been supported by CNR under ATC contract. Most of the contents of this paper were presented in ref. 8.

References

1. P. R. Drouilhet, "DABS: a system description", *MIT Lincoln Lab., Rep. ATC* 42, November 1974.
2. E. J. Kelly, "DABS channel management", *MIT Lincoln Lab., Rep. ATC* 43, *January* 1975.
3. *E. W. Dijkstra, "Cooperating sequential processes."* p. 43–112 In "Programming Languages" (F. Genys, ed.). London: Academic Press, 1968.
4. S. Lauesen, "A large semaphore based operating system", *Comm. ACM*, **18**, No.7, p. 377–389, (1975).
5. G. Bucci, D. Maio and P. Tiberio, "Struttura di un sistema per la gestione in tempo reale del canale di un radar secondario ad indirizzo discreto", *Proc. of LXXXIX AEI Conf., Catanzaro, September* 1978.

6. M. Giubbolini, D. Maio and G. Roselli, "La gestione del canale di un radar secondario ad indirizzo discreto in un sistema di controllo del traffico aereo", *Rep. CIOC-CNR No.* 40, June 1978.
7. A. Spiridion and A. D. Kaminsky, "Interrogation scheduling algorithms for a discrete address beacon system", *MIT Lincoln Lab., Rep. ATC* 19, October 1973.
8. G. Bucci, D. Maio and G. Roselli, "An interrogation–reply scheduling technique for a discrete address beacon system operating under low air traffic and strict computer time constraints", *Proc. Int. Conf. on Digital Signal Processing, Florence, September* 1978.

4.4 Other applications

Arrival Time Determination in Explosion Seismology

DRISS ABOUTAJDINE, ZINE EL ABIDINE AMRI, MOHAMED NAJIM, and JACK-GERARD POSTAIRE

Laboratoire d'Electronique et d'Etude des Systèmes Automatiques, Faculté des Sciences, Rabat, Morocco

In explosion seismology, the identification and measurement of arrival times of reflected energy is the first and most important process in relating seismograms to subsurface structure. A scheme to automatically detect the reflected waves and compute their arrival times is designed and tested. It is essentially based on the analysis of the variations of the signal power concommitent with the recording of the seismic events.

Introduction

Much of our knowledge of the structure of the continental crust has been obtained by explosion seismology. The basic method is to determine the travel time of seismic waves between a shot point and several seismic recorders at the surface.[1] The energy is released by the explosion of small charges of dynamite rings through the undergound. These seismic waves are partially reflected back to the surface at the interfaces where the acoustical properties of the rock change. The determination of the differences between the time of the shot and the arrival times of the seismic waves at a series of seismometers in a straight line provides a means for mapping of subsurface geological structures.[2]

In such experiments, seismometers with three component sensors are located along a profile of about 300 km. The electrical outputs are amplified and recorded in an analogeous form simultaneously with a timing signal. For an array of 24 seismometers, one experiment represents a total of 96 signals which must be analysed to extract significant information. On each record, the analyst must recognize the successive arrivals of the different waves. Some of them will have travelled direct from the shot point to the recorder through the upper layer of the crust; others are transmitted for most of their path by the underlying mantle.

The arrival time and the amplitude of the different waves depend on the as yet unknown configuration of the crust and, of course, on the distance between the shot point and the recording stations. Hence, the analyst has to recognize several waves of different amplitudes arriving at various times, and this must be done on a large number of records.[3]

To reduce this heavy demand from work which needs an expert's attention, an algorithm for automatically determining the arrival times of the seismic waves has been developed for implementation on a digital computer. The algorithm is based on the analysis of the variations of the mean power received by the recording station during a fixed time interval preceding the observation time (see Section 1). Without any particular assumptions of the properties of the seismic signal, one can expect an increase in power when the head of a seismic wave reaches the seismometer. Hence the arrival of a wave can be located at the beginning point of an ascending phase of the signal power.

We propose a procedure for accurately detecting this point, whatever the magnitude of the variation of the signal power, based on an automatic recognition procedure which identifies the time intervals where the power has a tendency either to increase or decrease (Section 2). The total variations of the signals power in such intervals are then determined to detect any significant increase. The beginning of the time interval corresponding to such an increment indicates the arrival time of the corresponding wave.

The procedure has been tested with actual field data. Some results are presented in the last section of this paper. They show that, without any frequency filtering, the proposed method is well suited to recognition of the arrival of different seismic waves, even for noisy signals.

1. Determination of the power variations

The proposed procedure is based on the analysis of the variations of the power involved in the ground's response to incoming reflected energy. When the head of a wave reaches a seisometer, the amplitude of the elastic disturbances is recorded by means of three orthogonal components. Two of them ($x(t)$, $y(t)$) describe the movements of the ground in the horizontal plane while the third one ($z(t)$) indicates the vertical displacements of the ground.

The mean power involved in the movements of the ground between time $(t-T)$ and time t is proportional to $P(T, t)$:

$$P(T, t)=\frac{1}{T}\int_{t-T}^{t}(x^2(\tau)+y^2(\tau)+z^2(\tau))\mathrm{d}\tau \tag{1}$$

When T is fixed, the time function $P(T, t)$ represents the mean power received by the seismometer during the time interval T preceding the observation time t.

If at time t, the head of a wave reaches the recording station, then $P(T, t)$ begins to increase. This increasing phase is followed by a decreasing one which corresponds to the amplitude attenuation of the ground's movements succeeding the passage of the head of the wave. Figure 1 shows the variations of the mean power $P(T, t)$ involved in only the vertical displacement $z(t)$ of a seismometer. In this case $P(T, t)$ reduces to:

$$P(T, t) = \frac{1}{T} \int_{t-T}^{t} z^2(\tau) \mathrm{d}\tau$$

The function $P(T, t)$ is obtained by computing the average energy of the incoming signals during a time interval T. $P(T, t)$ can then be viewed as a smoothed version of the instantaneous power of the signals. Consequently, T must not be chosen too large, otherwise the sensitivity of the method could be very low and small waves could escape detection. On the other hand, experiments show that, provided T is large enough, the mean power of the noise computed within a sliding window of width T is approximatively constant along the seismograms.

The analyst must then find a compromise so that good filtering of the effects of the noise combined with good sensitivity ensure the effectiveness of

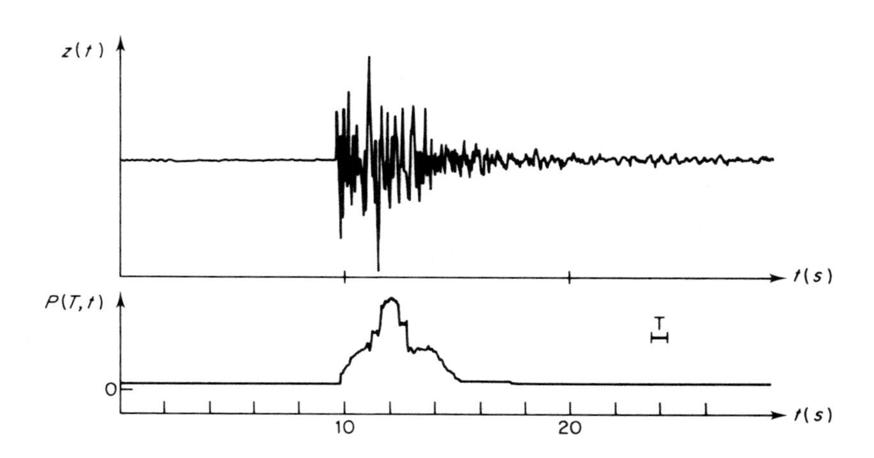

Fig. 1. The function $P(T, t)$ versus t for a one-dimensional displacement $z(t)$ of a seismometer.

the procedure. The choice of the parameter T, which is of primary importance, will be discussed in Section 3.

Before proceeding, let us consider the computational aspects of the determination of the function $P(T, t)$. First of all, the three signals $x(t)$, $y(t)$ and $z(t)$ are simultaneously digitized with a sampling period θ equal to 10^{-2} seconds. The discretized signals $x(i)$, $y(i)$ and $z(i)$ are then squared. Finally they are added up so that the integral which defines $P(T, t)$ can be easily computed in the discrete form.

$$P^*(T, n) = \frac{1}{m} \sum_{i=(n-m)}^{i=n} (x^2(i) + y^2(i) + z^2(i)) \qquad (n = 1, \ldots, N) \tag{2}$$

provided the time interval T is equal to an integer number m of sampling periods θ. The arrival time of the seismic waves can then be found from the determination of the time intervals for which the function $P^*(T, n)$ tends to increase. Indeed, the beginnings of the ascending phases of the power denote the arrival times of the heads of the waves at the seismometer.

Unfortunately, when the signals are corrupted by noise, the ascending phases of $P(T, t)$ are not always monotonic. (cf. Fig. 1). A procedure which ignores accidental variations and recognizes the *tendency* of the power to increase or decrease is developed in the next section.

2. Analysis of power variations

The ascending and descending intervals characterizing the variations of the power must now be defined and detected from the discrete function $P^*(T, n)$. First, let us consider the sequence of finite differences such that

$$\Delta P(n) = P^*(T, n) - P^*(T, n-1) \qquad n = 1, \ldots, N+1 \tag{3}$$

with

$$P^*(T, 0) = 0 \quad \text{and} \quad P^*(T, N+1) = 0$$

Figures 2(*a*) and (*b*) show a discrete function $P^*(T, n)$ and the corresponding finite difference plot.

Ascending and descending intervals are defined as positive and negative strings of finite difference values. If p_j indicates the position of the jth sign change, the magnitude of the ascending and descending intervals, denoted by Q_j, can be defined as follows:

$$Q_j = \sum_{n=p_{j-1}}^{p_j - 1} \Delta P(n) \qquad j = 1, 2, \ldots, S$$

where S denotes the number of sign changes. Q_1 is obtained from this definition if we assume that the first sign change occurs with $\Delta P(1)$, that is:

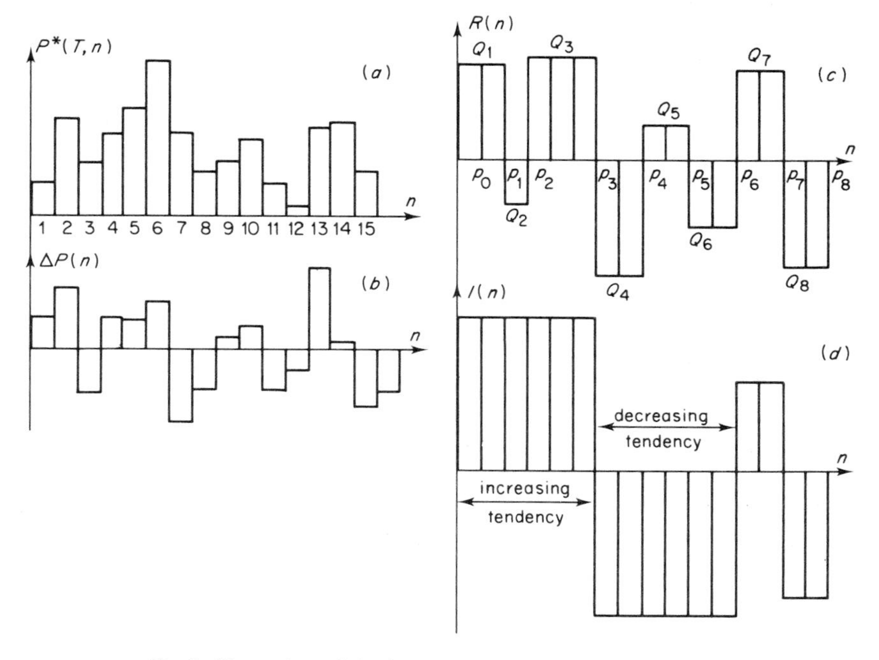

Fig. 2. Illustration of the four successive steps of the procedure.

$p_0 = 1$. The last Q_j is obtained if we assume that a sign change follows the last sample of the sequence $\Delta P(n)$, that is: $p_S = N+2$.

The magnitudes of the intervals of monotonic variation can be plotted as a function $R(n)$ of the time, defined as

$$R(n) = Q_j \quad \text{for} \quad n \in [P_{j-1}, P_j]$$

Figure 2(*c*) shows this function $R(n)$ which indicates the ascending and descending intervals of the power as well as their magnitudes.

To recognize the tendency of the power to increase or decrease, the magnitudes of the intervals of monotonic variation must be compared. Whenever the magnitude of a descending interval is smaller than the magnitude of the preceding ascending interval, the resulting variation of the power is an increment. Similarly, when the magnitude of an ascending interval is smaller than the magnitude of the preceding descending interval, then the resulting variation of the power is a decrement. This simple rule provides the basis of the recognition process of the tendency of the power to increase or decrease.

Starting with $i = 1$, each positive value Q_{2i-1} is compared to the next negative value Q_{2i} by computation of the sum Q_i.

$$Q_i = Q_{2i-1} + Q_{2i} \tag{4}$$

As long as Q_i is positive, the tendency of the power is to increase. This tendency reverses when a negative term Q_k is encountered. Then, all the amplitudes Q_j such that $1 \leqq j < 2k$ are algebraically added to determine the total increment of power corresponding to the first interval of increasing tendency. This interval spreads from the positions (P_0) to $(P_{2k-1} - 1)$ included (cf. figure 2(*d*)).

Starting with $i = k$, each negative value Q_{2i} is then compared to the next positive value Q_{2i+1} by computation of the sum Q_i,: $Q_i = Q_{2i} + Q_{2i+1}$. (Note that this new sequence of tests changes the value of ΔQ_k.As long as ΔQ_i remains negative, the power tends to decrease. This tendency reverses when a positive term, Q_l, is encountered; indicating that a decrement is followed by an increment of larger magnitude.

Then, all the amplitudes Q_j such that $2k \leqq j \leqq 2l$ are algebraically added to determine the total decrement of power corresponding to the interval of decreasing tendency which ranges from P_{2k-1} to $(P_{2l} - 1)$ included. The next interval is now characterized by a tendency to increase and is determined by computation of equation (4) starting with $i = l + 1$ instead of $i = 1$. The procedure continues until all the record is partitioned into adjacent intervals of increasing and decreasing tendency.

Figure 2(*d*) shows these intervals with the corresponding magnitude of the power variations under the form of a signal $I(n)$. Each positive edge of this signal corresponds to the beginning of an ascending phase of the power and represents the arrival time of a seismic wave.

3. Experimental results and conclusion

The effectiveness of this process has been tested with a set of actual field data recorded in Morocco by the Seismology Department of the University of Rabat, Morocco. The only parameter to be adjusted when processing the data is the width T of the time interval used to compute the power of the signals. As already noted, this parameter controls the smoothing of the variations of the instantaneous power of the signals. This property is used to filter out the effects of the interfering noise on the power variations.

This noise is essentially composed of low frequency ground-roll noise[4] and is available on the records before the arrival of the seismic events. This allows adjustment of the width T of the time interval in an interactive manner: T is taken equal to the width of the smaller sliding window such that the variations of $P(T, t)$ as a function of time t remain negligible.

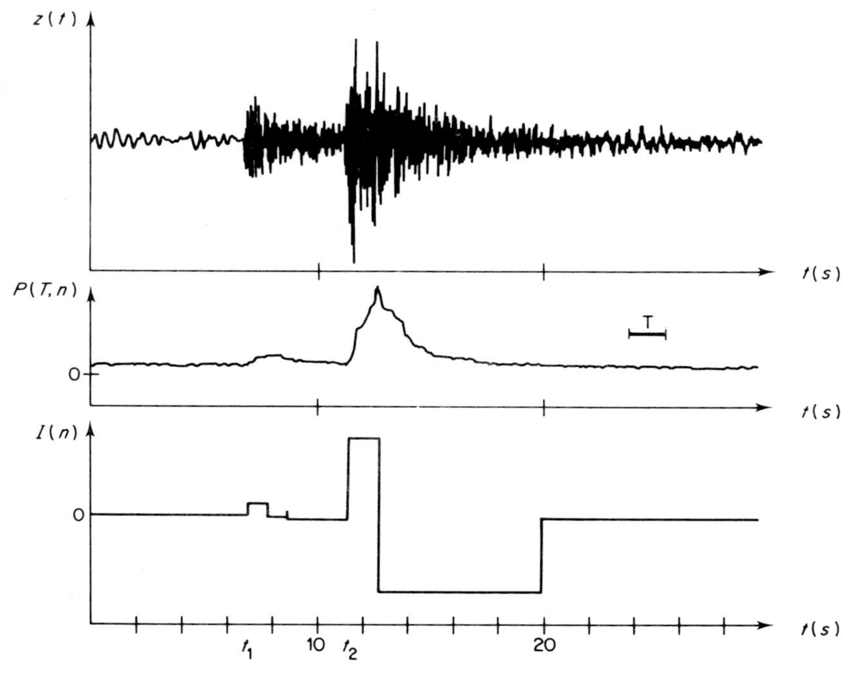

Fig. 3. Processing of a seismic record from a Moroccan area.

Indeed, as already mentioned, the experiments show that, for T large enough, the variations in the power of the noise are bounded within narrow limits. Hence, in these conditions, it is reasonable to assume that the majority of the events detected are genuine.

Figure 3 shows some typical results obtained with field recorded data. The top trace is the vertical component of the recorded events. The next trace represents the variations of the power of the signals computed with a time interval T of 2·4s. The result of the processing is summarized in the bottom trace on which two arrival times are clearly discernible. Time t_1 corresponds to the arrival of P waves and time t_2 to the arrival of S waves. This last trace replaces the original data, which is of a complicated nature, by a very simple representation of the seismic events.

This new representation contains only two descriptive characteristics of the seismograms which are of primary importance in determining the geological structure of an area. The first is the arrival time of the different reflected waves determined by the position of the positive edges of the signal $I(n)$; the second is the amplitude of the power variation corresponding to each event. The arrival times of the reflected energy can be automatically computed by a very simple analysis of the signal $I(n)$. This procedure provides the basis of the automation of the conventional technique which requires visual examination of the seismograms.

Additional work is being undertaken to apply this analysis to the processing of large arrays of sensors. The benefit to the analyst drawing the time-distance graph with the help of this procedure is uncertain at present, but it is reasonable to expect that speed will be increased, the role of the analyst being limited to the visual verification of the validity of the detections.

Acknowledgments

The authors wish to thank Mr. Driss Bensari and Mr. Michel Frogneux of the Seismology Department of the University of Rabat, Morocco, for helpful discussions and for providing the original data.

References

1. M. H. P. Bott, "The Interior of the Earth". London. Arnold 1971.
2. Smith, Steinhart and Aldrich, "The Earth Beneath the Continents" American Geophysical Union, 1966.
3. Coulomb and Jobert, "Traité de Geophysique Interne" Paris: Masson, 1973.
4. A. Robinson, "Statistical Communication and Detection". London: Hafner Publishing, 1977

Part 5
ART PROCESSING

Edge Extraction Techniques for Analysis of Art Works

M. BERNABÒ, V. CAPPELLINI and M. FONDELLI

Facoltà di Ingegneria, Università degli Studi Florence, Italy

1. Introduction

Automatic extraction by a digital computer of the geometrical elements and measurements from original works of art or photographs can assure a fast and objective analysis and facilitate the subsequent reconstruction of the formative processes.[1] In particular the extraction by the computer of the edges (contours, boundaries, lines) of works of art can give an immediate two-dimensional representation of the geometrical elements on a plotter, from which the three-dimensional construction of the work can be obtained through suitable computer programs. Owing to the objective and precise way of processing, complex curves can be extracted and shown in the two-dimensional or three-dimensional representation: these curves can then be described or approximated in an accurate way by mathematical relations.

For this purpose we have studied and applied different methods and techniques for automatic edge extraction using computers.

2. Digital edge extraction techniques and experimental results

Several methods and techniques have been studied for digital edge extraction. We have selected some of these techniques for automatic edge extraction from works of art by digital computer. These techniques and related algorithms can be divided in three main classes:

(a) two-dimensional (2-D) frequency filtering;
(b) grey-scale manipulation;
(c) local operators, based on heuristic methods.

2.1. Use of 2-D digital filters

A 2-D digital filter can be considered as a linear operator which processes an input sampled image $f(n_1, n_2)$ to give an output image $g(n_1, n_2)$ through an algorithm of the following type:[2-5]

$$g(n_1, n_2) = \sum_{k_1=0}^{N_1-1} \sum_{k_2=0}^{N_2-1} a(k_1, k_2) f(n_1 - k_1, n_2 - k_2) - \sum_{\substack{k_1=0 \\ k_1+k_2 \neq 0}}^{M_1-1} \sum_{k_2=0}^{M_2-1} b(k_1, k_2) g(n_1 - k_1, n_2 - k_2)$$

where $a(k_1, k_2)$ and $b(k_1, k_2)$ are the coefficients defining the digital filter and N_1, N_2, M_1, M_2 are suitable integers. If all $b(k_1, k_2)$ coefficients are zero, the 2-D digital filter is said to be of nonrecursive type or transversal, while if at least one $b(k_1, k_2)$ coefficient is different from zero the digital filter is said to be of recursive type.

For edge extraction, high-pass or bandpass 2-D digital filters are to be used for enhancement on grey-level variations.

2.2. Grey-scale manipulation

With grey-scale manipulation, the aim is to redistribute the original grey levels with the aim of enhancing the levels that are closely correlated to the edges, contours or boundaries. In this class of operations, we can include thresholding, by means of which we can represent only values greater or less than a given selected level (threshold).[6] Alternatively we can use a double threshold, representing only the values between two given levels which are closely correlated to the contours.

A similar procedure is used in the grey-level modification performed to make the distribution of the grey scale as homogeneous as possible.

2.3. Local operators

Local operators detect edges and contours by extracting gradients through a test on a given point and its close values. If the magnitude and direction of the gradient in a point are known and if the magnitude is greater than a given threshold, it is assumed that at that point there is an edge or contour whose direction is orthogonal to the gradient direction.

These methods can be divided in two groups.[6,7] To the first group belong the operators which evaluate two orthogonal components D_x and D_y of the gradient at each point and then derive its magnitude from

$$D = \sqrt{D_x^2 + D_y^2}$$

and its direction from

$$\phi = \tan^{-1}(D_y/D_x)$$

The evaluation of the gradient by considering matrices of 3×3 elements near the chosen point (i, j) is sufficiently accurate while being economic in computing time. Several methods are known along these lines with different weighting of the surrounding values. The most used masks are

(a) smoothed gradient masks:

$$D_x = \begin{bmatrix} -1 & 0 & 1 \\ -1 & 0 & 1 \\ -1 & 0 & 1 \end{bmatrix} \qquad D_y = \begin{bmatrix} 1 & 1 & 1 \\ 0 & 0 & 0 \\ -1 & -1 & -1 \end{bmatrix}$$

(b) Sobel gradient masks

$$D_x = \begin{bmatrix} -1 & 0 & 1 \\ -2 & 0 & 2 \\ -1 & 0 & 1 \end{bmatrix} \qquad D_y = \begin{bmatrix} 1 & 2 & 1 \\ 0 & 0 & 0 \\ -1 & -2 & -1 \end{bmatrix}$$

(c) isotropic gradient masks

$$D_x = \begin{bmatrix} -1 & 0 & 1 \\ -\sqrt{2} & 0 & \sqrt{2} \\ -1 & 0 & 1 \end{bmatrix} \qquad D_y = \begin{bmatrix} 1 & \sqrt{2} & 1 \\ 0 & 0 & 0 \\ -1 & -\sqrt{2} & -1 \end{bmatrix}$$

The equations for these masks can be expressed in the form

$$D_x(i, j) = f(i-1, j+1) + w \cdot f(i, j+1) + f(i+1, j+1) - f(i-1, j-1) - w \cdot f(i, j-1) - f(i+1, j-1)$$

$$D_y(i, j) = f(i-1, j-1) + w \cdot f(i-1, j) + f(i-1, j+1) - f(i+1, j-1) - w \cdot f(i+1, j) - f(i+1, j+1)$$

where the weight w can assume the values 1, 2, $\sqrt{2}$.

Some of the results obtained with these methods are presented. In Fig. 1 a photograph of the Brunelleschi Dome of S. Maria del Fiore in Florence is shown, while in Fig. 2 the results obtained by applying these methods to the original image† are illustrated. Similarly Fig. 3 shows the "Piazza Urbana" by a Florentine artist (1475–1485) from Santa Chiara of Urbino, while Fig. 4 shows the result of the processing of the red colour component of a detail of the previous image by one of the methods described.

The second group of algorithms for the evaluation of the gradient and then for the identification of edges is based on gradient detection by means of a set of templates or masks of different orientation, searching sequentially at each point for the best match between image sub-area and masks. Every mask of the set is superimposed on each sample of the image and the addition of products between the mask and the underlaying samples of the

† Digitized by Image Processing Institute (University of Southern California).

Fig. 1. Photograph of the Brunelleschi Dome S. Maria del Fiore in Florence.

image is performed as in the group of local operators. A gradient is assumed to be detected by the mask which gives the greatest value for sum of products and its direction is assigned to the direction of the mask. Each set of masks is composed of eight different 3×3 circular permutations of its elements around the central one.

The sets of masks more frequently used are obtained through a permutation of the following:

(a) Prewitt mask[8]

$$\begin{bmatrix} 1 & 1 & 1 \\ 1 & -2 & 1 \\ -1 & -1 & -1 \end{bmatrix}$$

(b) Kirsch mask[9]

$$\begin{bmatrix} 5 & 5 & 5 \\ -3 & 0 & -3 \\ -3 & -3 & -3 \end{bmatrix}$$

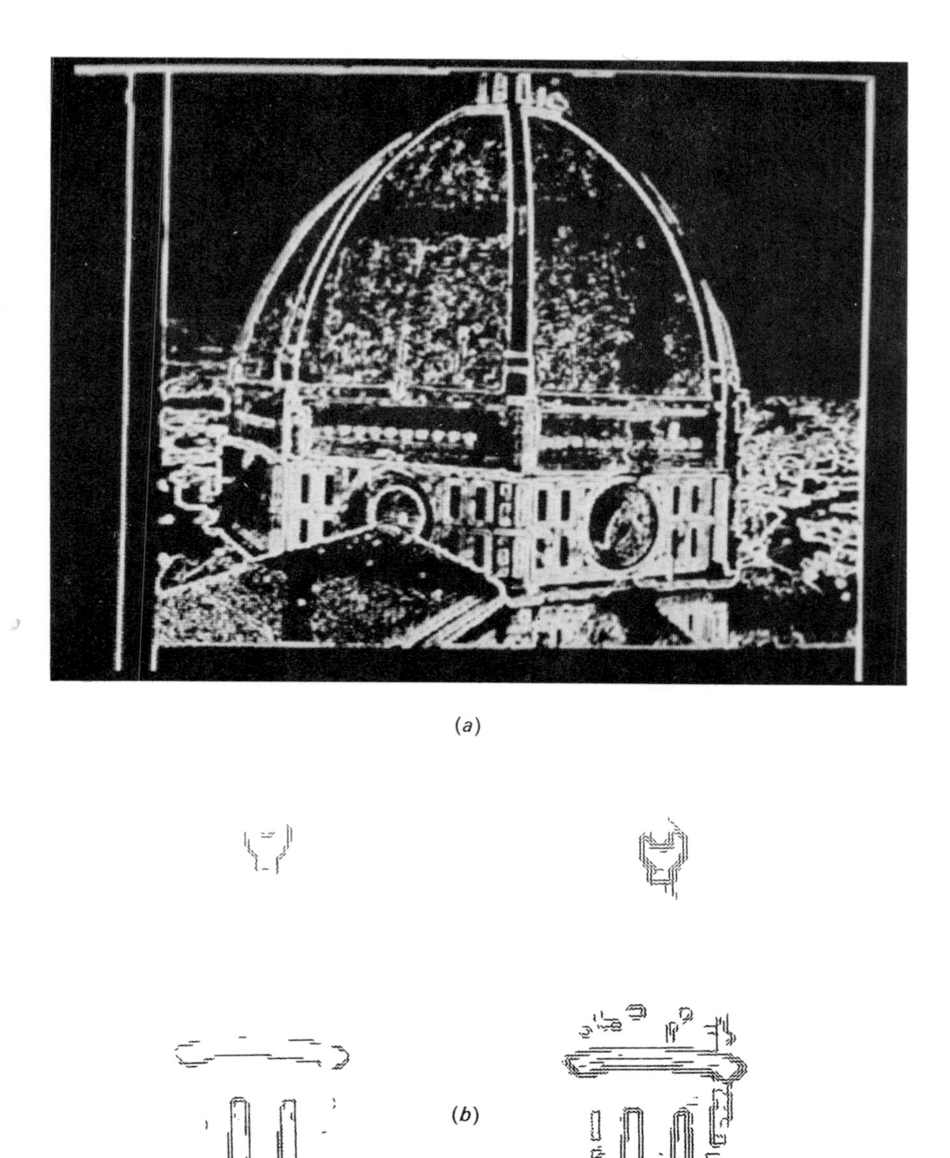

Fig. 2. (*a*) Result of processing an image as in Fig. 1 by means of a gradient mask (in cooperation with TESAK S.p.A. Florence). (*b*) Results of processing a detail of the image of Fig. 1 by means of the Sobel gradient algorithm with two different values of the threshold.

Fig. 3. "Piazza Urbana" by a Florentine artist (1475–1485) from Santa Chiara of Urbino.

Fig. 4. Result of processing a detail of the image of Fig. 3 by means of the Sobel gradient.

(c) Robinson mask[10]

$$\begin{bmatrix} 1 & 1 & 1 \\ 0 & 0 & 0 \\ -1 & -1 & -1 \end{bmatrix}$$

Chen and Frei[11] have improved the previous methods in order to define a deeper mathematical background to the various heuristic methods and they proved that it is possible to choose a set of nine masks which form an orthogonal basis spanning the nine-dimensional space whose elements are the 3×3 sub-areas of the image.

3. Comparison of efficiencies of different edge extraction techniques

Some comparison of efficiency can be made among the different edge extraction techniques described. In these comparisons two main aspects must be considered: the power and quality of the method or algorithm giving edge extraction; the complexity and cost (as computing time) that the method implies in its application.

Spatial filtering, may be seen to be very general and flexible, independent of particular problems, with a well established mathematical basis. For edge extraction in general high-pass digital filters are applied, as noted above. These can have two drawbacks: they do not perform very well with noisy images, because they enhance high-frequency components; and they suppress low-frequency components which are connected with the background of the image.

Grey-scale manipulation is useful in connection with other methods. In particular the threshold operator is very suitable for choosing the elements belonging to an edge and eliminating nonessential aspects of the image.

The methods described above using masks are based only on heuristic considerations, but are nevertheless often very useful due to their relative simplicity. The smoothed and Sobel gradient have the advantage of having only integer values in their masks, so that only integer multiplications are involved, with associated computational cost-saving. The isotropic gradient, however, has the property that the weights are proportional to their distance from the centre of the mask.

Some improvements on the other techniques can be achieved with the Chen–Frei algorithm,[11] which can more clearly detect the edges and with a better enhancement, at the price of only a little increase in the computational cost.[12]

Finally it is interesting to observe that good edge extraction results can be obtained by applying twice one of the methods described.

Acknowledgment

We thank Dr. G. Benelli, Ing. L. Baroncelli and Ing. A. Del Bimbo for their useful cooperation.

References

1. C. L. Ragghianti, "Arte: fare e vedere". Firenze. Vallecchi, 1974.
2. V. Cappellini, A. G. Constantinides and P. L. Emiliani: "Digital Filters and Their Applications". London: Academic Press, 1978.
3. V. Cappellini and P. L. Emiliani, Design of Some Digital Filters with application to Signal and Image Processing". *Proceedings of Summer School on Circuit Theory Praha, September* 1974. **1**, pp. 161–176.
4. M. Calzini, V. Cappellini and P. L. Emiliani, "Alcuni filtri numerici bidimensionali con risposta impulsiva finita". *Alta Frequenza*, **44**, No. 12, pp. 747–753 (Dec. 1975).

5. M. Bernabò, V. Cappellini and P. L. Emiliani, "Design of 2-dimensional recursive digital filters having a circular symmetry." *Electronics Letters*, **12**, No. 11 (May 1976).
6. P. Kammenos, "Operateurs locaux pour l'extraction des contours". *Rapport L.T.S*. 76.05, Laboratoire de Traitement de Signaux, Lausanne, 1976.
7. B. Bullock, "The Performance of Edge Operators on Images with Textures". Hughes Aircraft Company Research Laboratories, Techn. Rep., Oct. 1974.
8. B. S. Lipkin and A. S. Rosenfeld, "Picture Processing and Psychopictorics". London: Academic Press, 1970.
9. R. Kirsch, "Computer determination of the constituent structure of biological images". *Comput. Biomedic. Res.*, **4** (1971).
10. G. S. Robinson, "Detection and Coding of Edges Using Directional Masks". University of Southern California, Los Angeles, *USC–IPL, Rep.* 660, *March* 1976.
11. W. Frei and C. C. Chen, "Fast boundary detection: a generalization and a new algorithm". *IEEE Trans. on Computers*, **C-26**, No. 10, (Oct. 1977).
12. V. Cappellini, M. Fondelli and M. Bernabò: "Digital Edge Extraction on Works of Art". Sound Sonda, n.5, Università Internazionale dell'Arte, Firenze, 1978.

Analysis and Reconstruction of Art Works using a Digital Computer

R. BRUNO

Scuola Normale di Pisa, Pisa, Italy

and

L. CRISTIANI TESTI and T. ZANOBINI LEONI

Istituto di Storia dell'Arte, Università degli Studi di Pisa, Pisa, Italy

1.

The analytical system actually initiated in 1974 using an electronic computer is based on the criterion of equivalence which informs and regulates the conception and practice of perspective construction. The procedures elaborated on the basis of this criterion allow us to achieve planimetric restitution in orthogonal projection of one or more planimetric perspective sections of a work, proceeding from the equivalence of a geometric element depicted on the image projection plane (most frequently a trapezium) with a well known plane geometric figure with right angles (most frequently a square). By applying and developing this principle, the system also reproduces the perspective–volumetric elements and sections of the work under consideration in plan and in front elevation.

In the "School of Athens" by Raffaello, the system allowed: (1) The comparison of the equivalence of the flooring panels (trapeziums) of the proscenium composed of squares, and the equivalence of the bands including horizontal and transversal bands composed of rectangles or squares. (2) The reconstruction of the orthogonal planimetry of the perspective planimetric section composed of these elements. (3) The determination of the difference of levels, the depth, and the altimetric profile composed of the three steps of the stairway. (4) The reconstruction of (a) the essential vertical elements of the perspective of the building; (b) the planimetry of the proscenium placed on the second level of the flooring; (c) the ground work and the vertical section of the two shortened wings of the building and of the interposed central room. (5) The singling out and reconstruction on this basis of the ground work and the vertical section of the four solids corresponding to the volumetry of the architectural surroundings symmetrically distributed in relation to the crossed main axes and of the two solids corresponding to the volumetry of the transversal wings of the building.

The result allowed us to obtain perspective restitution, using the plotter, of the global composition of all the elements and of all the essential components of the work (Fig. 1). The automatic operation of perspective restitution verifies the correctness, congruence or non-arbitrariness of the entire process of analytic reconstruction and hence the general pertinence of the basic criterion from which the procedure derives.

The results obtained must be considered as experimental and therefore approximate. Experience has, in fact, clarified and detailed more precisely the system's intrinsic limits and has shown the need for elaborating a second system having more rigorous geometric reasoning and more direct application to the problem of analytic restitution of art works in perspective (not only monocentric) construction.

2.

The process of analysis and restitution of constructive processes corresponding to the exactness criteria mentioned for monocentric, bi- or tricentric space-perspectives is planned here as a conversational. The system is founded upon basic Euclidean logic and the relevant operating instructions and, in this sense, may be defined as procedural. Consequently a characteristic of the Generalprosp system, as here organized (when it meets the requisites of exact geometric construction, as well as being acceptable, congruent and adaptable), is its flexibility. The necessity for exact, though not necessarily unique, perspective representation clearly derives from the basis of the procedural system, which may be defined as the transformation of the coordinates of one point in relation to the object's space coordinate system, to the coordinates of its projections on the plane of the picture measured in the coordinate system of the picture. Obviously, in order to identify the position in the three-dimensional structure of a point projected on the plane of the picture, one must start with elements of definite construction and commensurability, from whose elements one can infer simultaneous relationships between the coordinates of the plane of the picture and those of the spatial construction.

The first practical step of the operation is establishing the vanishing point

Fig. 1. Plotter graphs showing successively hidden lines (CNUCE 1974).

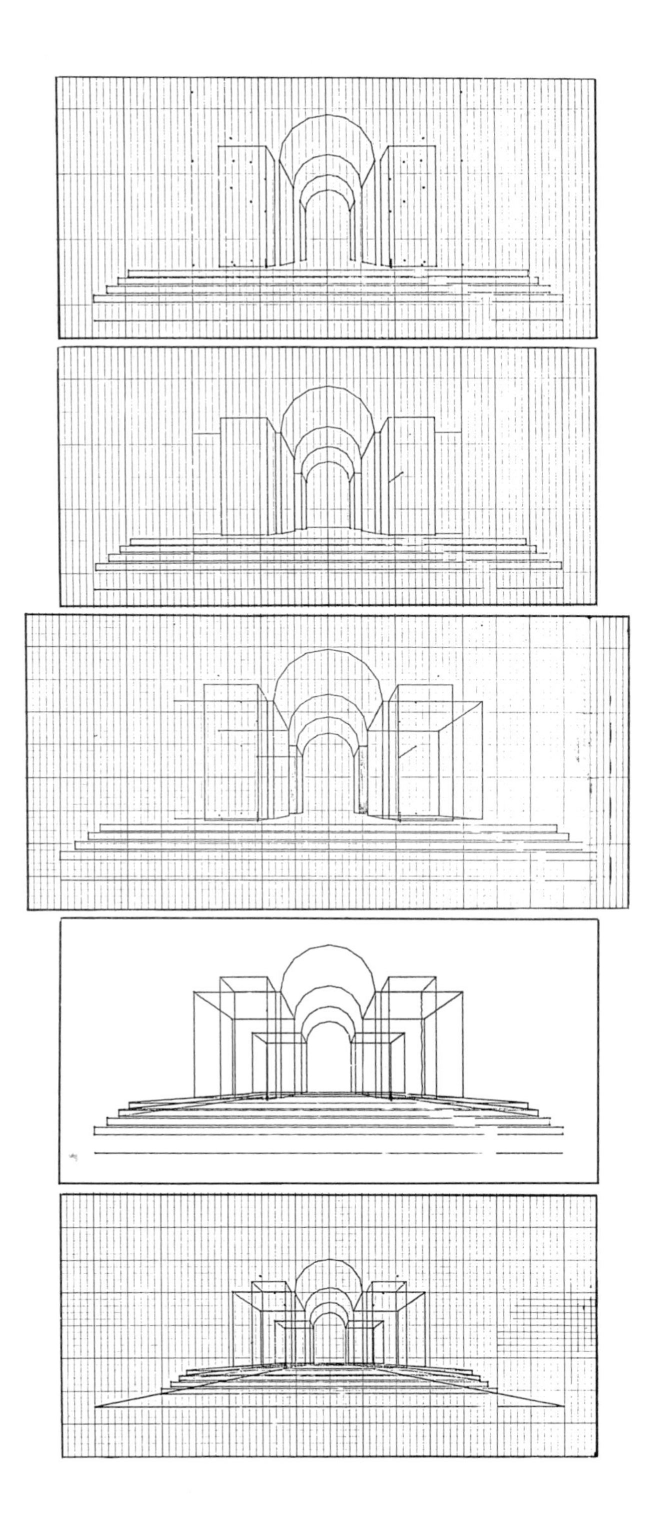

or points, from which the regularity or lack of it of the space-perspective construction is verified (Fig. 2(*a*)) and which is then numerically determined (Fig. 2(*c*)) in relation to the reference grid determined in the procedural system (Fig. 2(*b*)). These x, y coordinates, together with the coordinate system, define the coordinate system in object space. It is in this system that the relative position of the object points in the picture plane will be expressed.

The picture coordinate system is determined by choosing a suitable point as origin and defining three orthogonal directions. Obviously, the tendency is to select geometrically disposed elements, such as the corners of the squares in the pavement or flooring, a capital abacus, or column bases or ceiling squarings. The x, y coordinate system is frequently, but not necessarily, defined by taking the lower left corner of the picture as the origin and is called the coordinate system of the picture. We record the points in the picture to be reconstructed in this system. The method encounters serious problems in attempting to record scarcely discernible points and lines in photographic reproductions or in the original works, which may considerably distort the results. It is hoped that these difficulties will be overcome using a more objective and rapid operation involving digitization which electronically identifies and assigns the points. (The two coordinate systems turn out to be rotated with respect to one another, a feature which allows the recognition and solution of situations in which the figures are not parallel to the plane of intersection of the visual pyramid.)

The identification of the vanishing point involves the selection of two parallel lines and the numerical determination of the coordinates of four points which define them, two on one line, two on the other. An alternative method is the selection of a geometric element, for example a segment whose position is certain in the planimetric object system, with another adjacent segment whose represention is proportionately larger or smaller in perspective. These proportions are numerically determined and identified in terms of the coordinates of their defining points. (A further method is to read off the coordinates of the geometrically drawn vanishing point, as done here).

The second operation is assigning the internal coordinates, which necessarily follow from the first step. When there is only one vanishing point, as in the cases considered here, the procedure is to single out a line, in a known geometric element, which is oblique to the direction of the vanishing point.

Fig. 2. (*a*) Verification of any orthogonal line convergence in a vanishing point F. (*b*) Verification of proportional ratios on image plane. (*c*) Vanishing point F. Reference length; brick $AB = A_1B_1$. (*d*) The signed points are measured with respect to x and y axes. (*e*) Planimetric restitution of St. Sebastian. Viewing point = eight bricks.

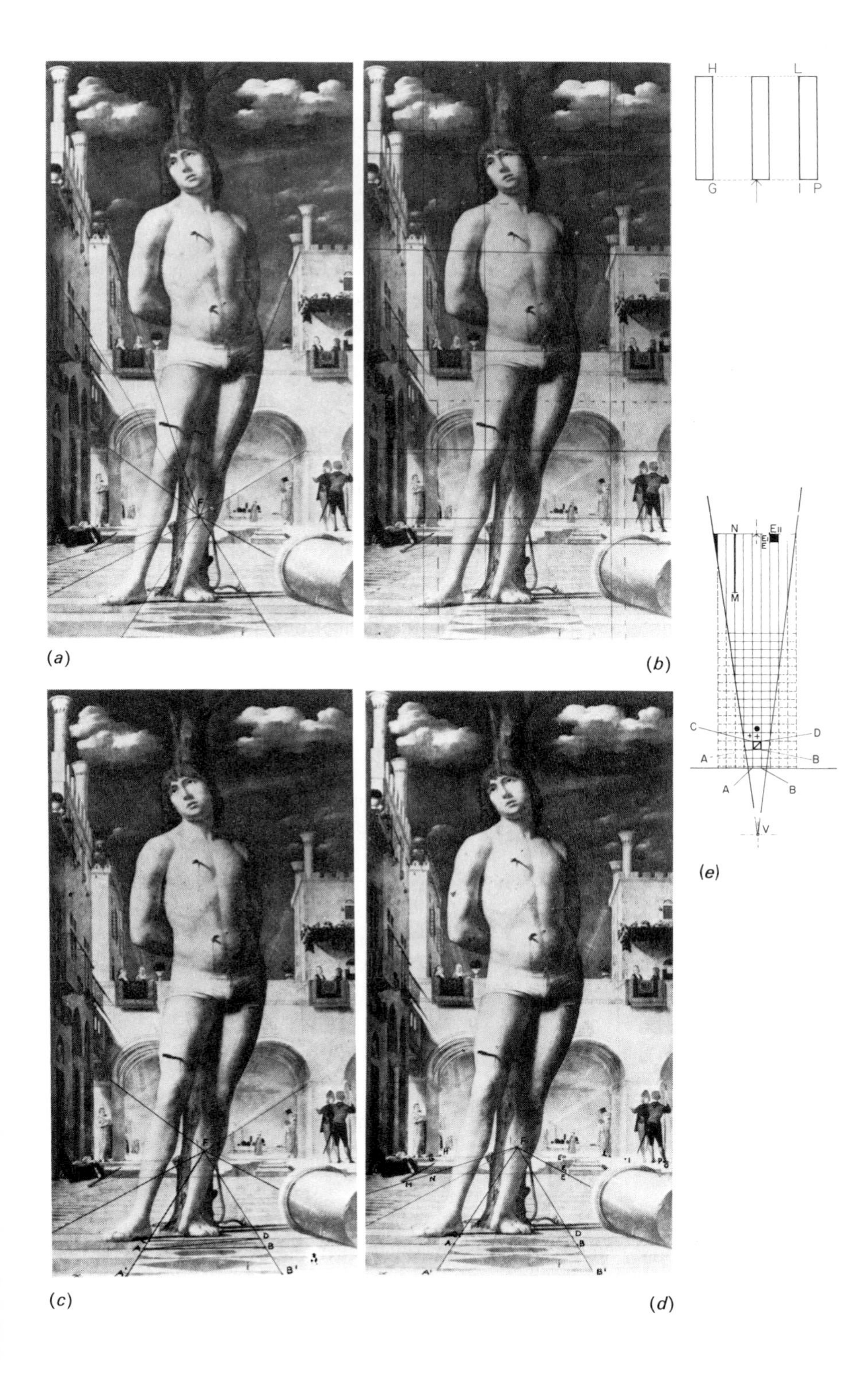

(a)
(b)
(c)
(d)
(e)
H
L
G
I
P
N
M
C
D
A
B
V
F
A'
B'

Having performed the first two steps, the third consists of identifying the orientation matrix (which may also be determined by calculating the angles between the space of the coordinate system of the picture, the edges of the picture and the axes of the object coordinate system).

We postpone discussion of the mathematical fine detail. What is of interest here is the demonstration of the method's fulfillment of the methodological criteria and its real advantages over the traditional geometrical-analytical operation as they are demonstrated in application to the St. Sebastian in Fig. 2.

From this analysis and subsequent operations of the analytic program we obtain the unforeshortened distances of points which lie on the same plane or on different planes. From this we derive restitution of the planimetric reality of the subject required to generate the perspective representation we see in relation to the vanishing points chosen by the artist and to the viewing-point, which also arises from the analytic program and operative procedure phases. As is evident from the above, we also obtain restitution of the front elevations in their true dimentions and, hence, vertical sections perpendicular or parallel to the picture plane. As a result we can also construct alternative, but valid views of the space-perspective representation used by the artist.

3.

The painting "La Flagellazione di Urbino" of Piero dei Franceschi was chosen for its graphic complexity and rigorous perspective, and because of the variety of solutions produced by numerous workers who have attempted restitution of the plan.

The first task is recording data for processing. Since single readings are unreliable, we adopted a method used in topography based on the optimization of numerous measurements of the same object. Once the system of reference axes for assigning the coordinates of points in the picture was established, the data were processed to give results diverging for each element by about 1 mm (Fig. 3.). We were able to obtain the following coordinates for the viewing-point (with one working photo of $36{\cdot}5 \times 25{\cdot}5$ cm).

$$x_v = 18{\cdot}35 \pm 0{\cdot}0947 \quad \text{(average shifting of the picture)}$$

$$y_v = 7{\cdot}204 \pm 0{\cdot}070 \quad \text{(average shifting of the picture)}$$

Fig. 3. (*a*) Reference point in the first phase of the detection. (*b*) Determination of the middle point V and of error field. (*c*) Geometrical scheme for the determination of the true distance of a reference point P from the painting.

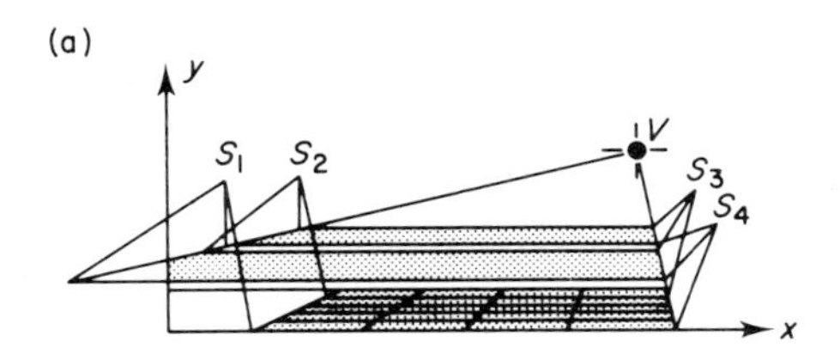
(a)
y
V
S1
S2
S3
S4
x

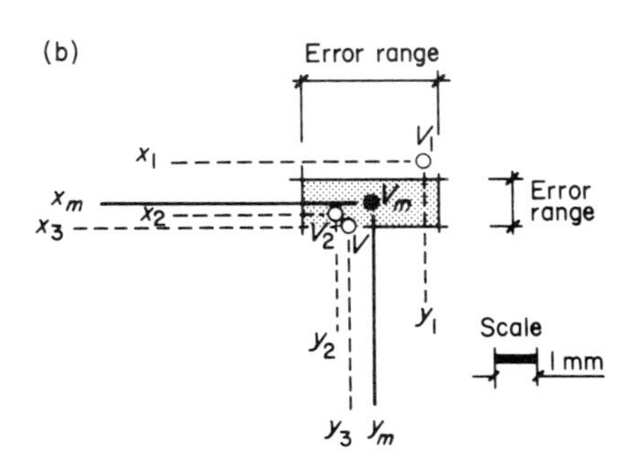
(b)
Error range
x1
V1
xm
Vm
x2
x3
V2
V
Error
range
y1
y2
Scale
1 mm
y3
ym

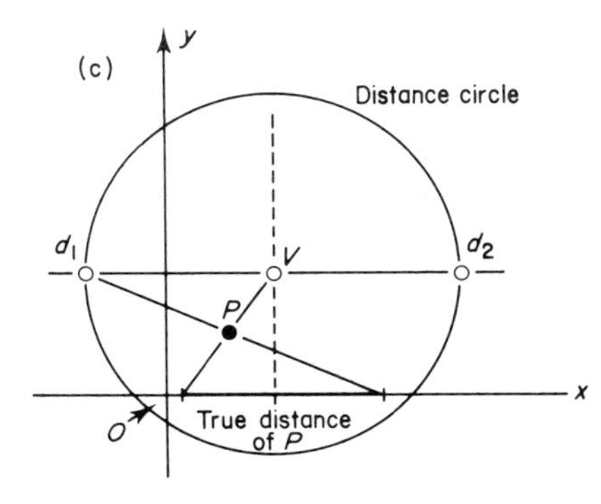
(c)
y
Distance circle
d1
V
d2
P
x
O
True distance
of P

In establishing the distance between the spectator and the picture, the margin of error increases because of longer processing and inherent errors in the artist's drawing.

The results vary by about 2 cm:

$$D = 67{\cdot}58 \pm 1{\cdot}99 \text{ cm}.$$

On the basis of these two parameters (viewing-point and distance), we can deduce the positions in the reconstruction of all the major points of a picture, and these can be displayed, if wished, by means of a plotter or video.

Application of Digital Image Processing to Archaeological Prospecting

V. CAPPELLINI and M. FONDELLI

Facoltà di Ingegneria, Università degli Studi, Florence, Italy

1. Introduction

Aerial photography is a very useful technique for archaeological research and prospecting. Works at 2–3 meters below ground level or at 20 meters below sea level can be detected and recognized by using suitable photographic techniques. Certain photographic details, such as shadow-sites, grass-weld-cropmarks, soil-marks, damp-marks are of particular impor-

Fig. 1. Part of an aerial photograph showing the site of Capaccio Vecchio (authorization n. 12, 10 Jan 1975).

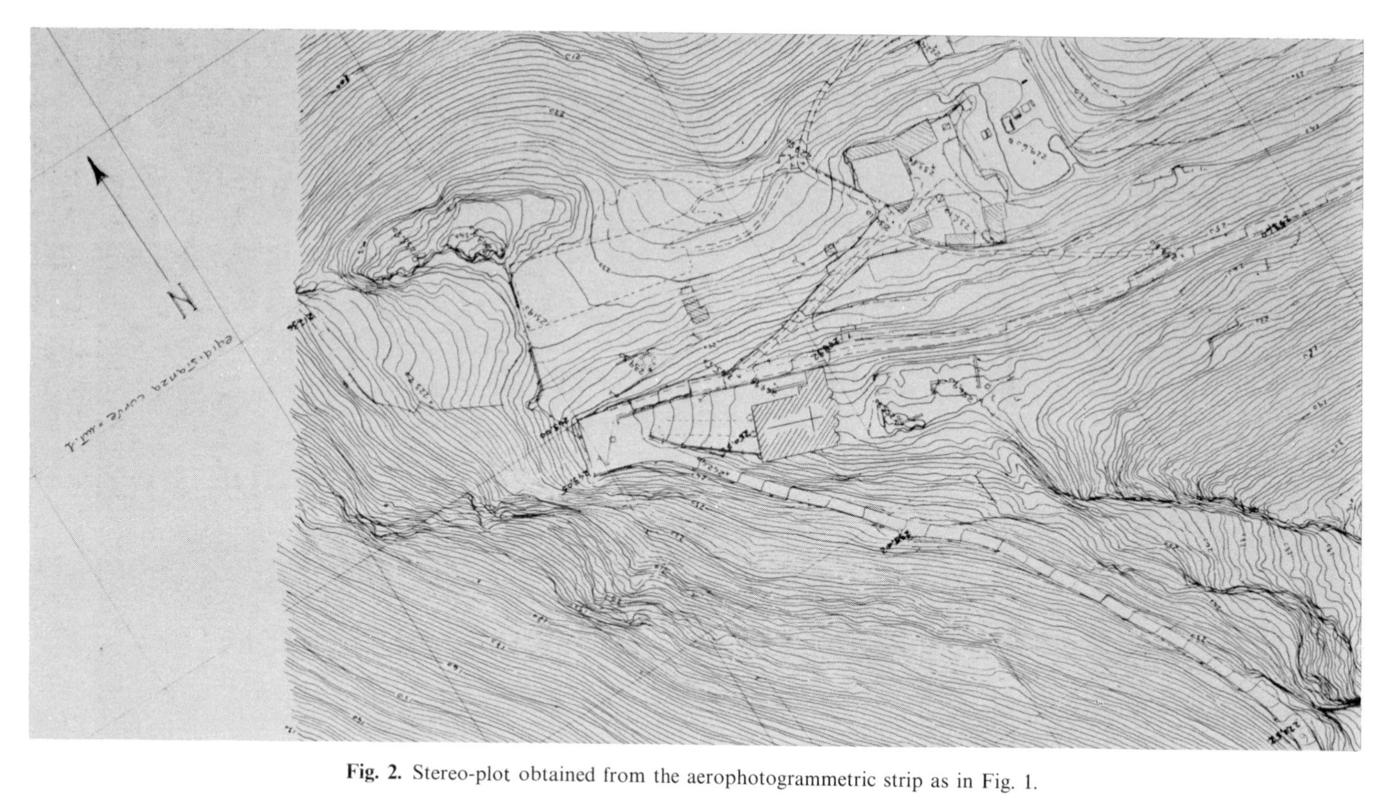

Fig. 2. Stereo-plot obtained from the aerophotogrammetric strip as in Fig. 1.

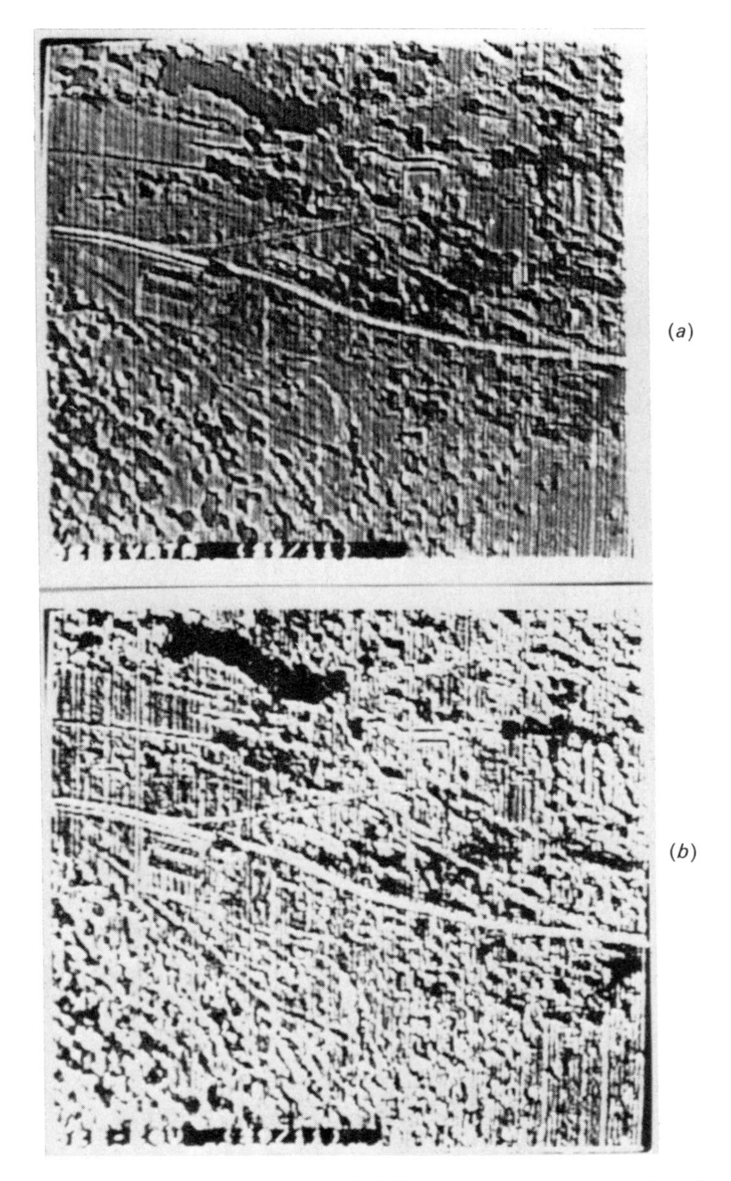

Fig. 3. Results of application of a derivative digital filter (*a*) and a derivative digital filter with equalization (*b*).

tance. In addition the physical conditions (temperature, humidity, etc.) in which the photographs are taken must be considered carefully. Thus, analysis of a photographic image for archaeological purposes is similar to a decoding operation, involving pattern recognition and decision steps.

Digital processing techniques can be very useful in solving, at least partly, such problems and identifying ancient works. In particular digital filtering, edge extraction and pattern recognition techniques are of considerable interest. In the following some examples are given of these techniques.

2. Examples of digital processing of aerial photographs in archaeological research

Figure 1 shows a part of an aerial photograph, taken by the Istituto Geografico Militare in Florence for the Cassa per il Mezzogiorno (sheet 198 of the Italian Map) of the site of Capaccio Vecchio. Figure 2 shows the stereo-plot obtained from the aerophotogrammetric strip (as shown in Fig. 1). Figure 3 shows the result of applying digital filters (in cooperation with Istituto di Elettrotecnica ed Elettronica dell'Università di Trieste): in (*a*) the result of a derivative digital filter is shown, and in (*b*) that obtained with a derivative filter with equalization.

Figure 4 shows the grey-level perspective, after some level digital manipu-

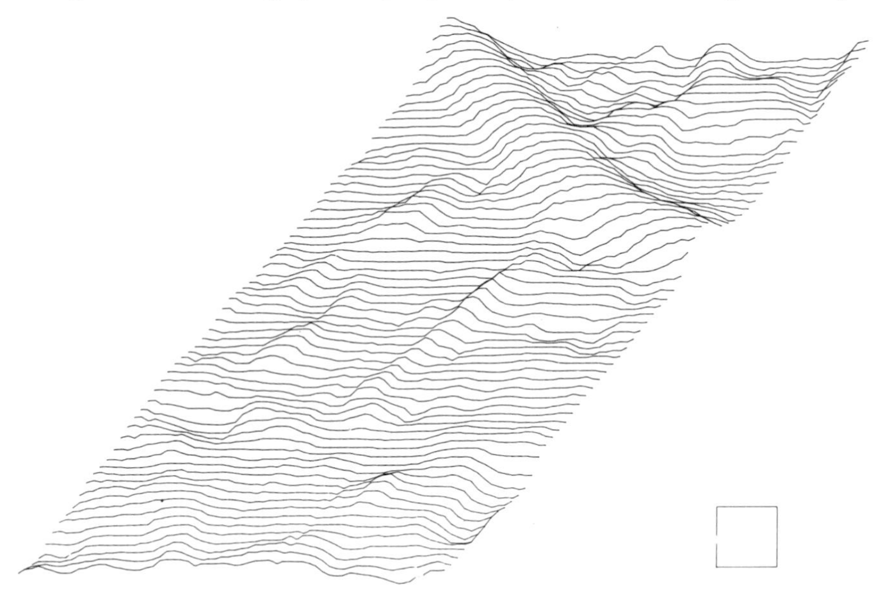

Fig. 4. Grey-level perspective, after some digital level manipulation, of a small part near the lower church in Fig. 1.

lation, of a small part (64 × 64 samples) near the lower church in Fig. 1, where the archaeological works are expected to be found: some clear alignments are recognizable.

Digital processing techniques can, as these examples indicate, be very useful in extracting information on archaeological works from aerial photographs by indicating alignments, grids and geometrical structures through filtering, enhancement and pattern recognition operations.

Subject Index

A

Accumulated processing time, 129, 131
Acoustics, 138
Acoustic wave propagation, 53
Active filters, 31
  message list, 242
Adaptive, filter, 21–25
  predictive coding, 145
  time domain equalizer, 25
  transversal LMS filter, 21, 25
Adder, two input, 128
Adjustable recursive filter, 21
Aerial photography, 279–283
Air traffic control,
  ATCRBS, system 239
  radar beacon, 239
  system 239–249
Aliasing, 147
Algorithm,
  back projection, 207–212
  Baver, 12–19
  Bohme, 84
  Chen–Frei, 268
  close fit, 244
  filtered back projection, 210
  FIFO, 248, 249
  Levinson, 35, 37, 44
  LMS, 21–25
  prime factor, 66
  Remez exchange, 3, 4, 8
  Rivard, 97, 101
  scheduling, 244
  time scheduling, 129
All-pass 20 phase filter, 51
Amplitude filter, 48
Analysis and reconstruction of art works, 271–278
Antenna beam, 240–241
Applications of algebraic numbers to convolution and DFT, 61–68
Archaeological prospecting, 249–283
Arithmetic table, 109
ARMA models, 39–40
ARMA modelling factor, 42
  modelling filters, 44
  transfer function, 44
Arrival time determination in explosion seismology, 251–258
Art works,
  use of edge extraction techniques, 261–269
  use of digital computers, 271–278
Ascending phase, 256
Autocorrelation function, 169
Autoregressive, approximation, 37
  model, 35
  moving average, 39
Azimuth,
  antenna, 241
  limit, 242
  order, 244–247

B

Back projection, 211–212
  algorithm, 207–212
Baver's algorithm, 12–18
Bessel functions, 209
  zero-order, 88
Bilinear transformation, 30–32, 42, 145
Binary numbers, 105
Bit stripping, 139
Bohme algorithm, 84
Boolean equation realization, 106–108
Boundary effects, 70
Branch filtering, 171

C

Causal filters, 47
CCITT, 147, 150, 159, 160
Chain matrix, 28, 31
Chain scattering matrix, 36, 37
Chain scattering factor, 40
Channel management, 240–242
  software functions, 242–243
Channel vocoder, 137, 139, 142, 145
Chebycheff tolerances, 175
Chen–Frei algorithm, 268
Chinese Remainder Theorem, 229–237
Circulant form, 79, 80
Circular permutations, 264
Circularly symmetric 2-D Fourier transforms, 87–95
Clipping, 148
Close-fit algorithm, 244
Coder
  13 bit, 148, 150–160
  14 bit, 148, 151, 154–160
Codes,
  companded, 152–154, 158, 159
  uniform, 152, 153, 158, 159
Companding order, 158
Compatibility, of linear and compander coders, 158, 159
Compensation of tissue absorption, 212–214
Compressor, 150
Computation time, 83
Computer simulation, stochastic signals, 161–168
Computerized emission tomographic system design, 207–216
Constant false alarm rate radar systems (CFAR), 222, 221, 218, 220
Convolutions, applications of algebraic numbers, 61–68
Convolutional filter, 210
Correlation function, 83
Cosine transform, discrete, 72
Critical path problem, 125, 129, 131
Cross talk, 154
Cyclic convolution, 66, 232, 233

D

Darlington recursion, 30, 40, 41
Data transmission, high-speed, 25
D.C. offset, 155–157
Dead zone, 148
Decimation in time, FFT, 62–64
Degree of flexibility, 131
Delay-free flow-graph, 125, 127
  loops, 27, 124
  network, 131
  time-weighted graph, 125, 127, 128
  time-weighted sub-graph, 128
Derivative digital filter, 282
Digital chirp filtering, 229–237
Digital computers, 271–278
Digital edge extraction, 261–269
Digital filters,
  derivative, 282
  design, 27–33
  frequency-dependent, 27–33
  lossless, 36
  representation, 124
  spectral factorization, 11–19
  topological considerations, 123–133
  using microprocessors, 123–133
  wave, 30, 36–38
Digital holography, 69
Digital interpolation of stochastic signals, 163–169
Digital processing systems, 105–115
Digital speech processing, 137
Digital switching, 147–161
Direct convolution, 210
Discrete Address Beacon System (DABS), 239–249
Discrete cosine transform, 72
Discrete Fourier transform (DFT), 9, 83, 84
  applications of algebraic numbers, to 61–68
  complex valued, 64
  direct shifted, 70
  shifted, 69–73
  standard, 70–73
  reverse shifted, 70
  2-D, 97
  Gauss–Gauss detection problem, 75
  time, 105
Distortion measure, 118
Doubly terminated filter design, 27
Double threshold, 262
Down link message, 239, 241, 242, 244

E

Edge extraction techniques for art works, 261–269
Elastic disturbances, 252
Embedded coding, 138
Emission tomographic system design, 207–216
Enhancement noise, 154, 157
Entropy, 117, 118
Equalizer adaptive time domain, 25
Equalization, 282
  linear phase, 52
  group delay, 53
Error probability, 217
Error, round-off, 17
Explosion seismology, 251–258

F

Factorization of polynomials, 11
  matrix, 11
Fast computation of Toeplitz forms, 75–85
Fast Fourier transform (FFT), 12, 79, 81, 83, 184, 191, 210, 213, 214, 229, 232, 233, 235, 237
  decimation in time, 62–64
  radix-2, 65, 67
  radix-3, 61–68, 101
  radix-4, 65
  2-D, 97–102
Feintuch's adaptive recursive LMS filter, 21, 22
Fermat numbers, 66
FIFO algorithm, 248, 249
Filtered back-projection algorithm, 210
Filtering, 210
  branch, 171–182
  strip techniques, 195
Filters,
  active, 31
  adaptive recursive LMS, 21–25
  ARMA modelling, 40, 44
  causal, 47
  design, 49, 50
  high order, 3–5
  half-plane fan, 54
  implementation, 176
  ladder, 27
  Lagrangian, 163
  minimum phase, 48, 55, 56
  orthogonal, 35
  otpimum, 164–167, 178
  phase, 48
    all-pass, 48
  recursive, 2-D, 47–50
    first order, 106
  2-D phase, 47–57
  transversal, 21, 25, 229
  wave digital, 30
Finite arithmetic structures, 105–115
Finite impulse response (FIR) filters, 11, 163
Finite and IIR complementary structures, 171–182
Finite complementary filters, 171–172
  linear phase, 3–9, 171
  low pass, 3, 163–169
  realization, 47
Finite polynomials, 110, 11
Finite state machine, 109
First order sensitivity, 22–24
Floating point multiplications, 127, 132
Flow procedure, 138
Formative processes, 261
Fourier–Bessel transform, 88
Fourier domain, 197–205
Fourier transform, 213
  application of 2-D, 87–95
  one-dimensional, 87–95
  monodimensional, 208
Fourier techniques in 3D reconstruction, 197–205
Frequency dependent linear transforms, 27–33
Frequency division multiplex, 171
Full-Ring packing efficiency, 244, 245, 247

G

Galois field, 108–115
Galois structures, 108
Gamma camera, 209
Gauss–Gauss detection problem, 75
Gaussian distribution, 156, 218
Gaussian integers, 67, 68
Gaussian noise, white, 168, 186
Gaussian process, 35, 77, 184, 186

Gaussian probability density function, 155
Gaussian random process, 155
Gaussian signal, 117
  source, 119
General prosp. system, 272
Gradient mask,
  isotropic, 263
  smoothed, 263
  Sobel, 263
Granular noise, 148, 151
Grey-scale manipulations, 261, 262, 268
Grid, rectangular, 90
Ground-roll noise, 256
Group-delay equalization, 53

H

Half-plane fan filter, 54
Hamming window, 168
Hankel transform, 88, 89
Hanning window, 94, 168
Hermitian conjugate, 35
Hexagon, 127, 128
Hexagonal area, 118
Hexagonally periodic sequences, 98, 101
Hexagonally sampled data, 97–102
High order filters, 3–5
High-pass inverse filter, 55
Hilbert transform, 12
Holography, digital, 69

I

Image data compression, 56
Image restoration, 184–185
Importance sampling techniques, 218, 220, 221
Impulse response, of FIR filters, 6
Independent identically distributed random variables, 117
  Gaussian source, 117
Inhomogeneous plane waves, 91, 92
Infinite impulse response (IIR) complementary filter, 172–176
  filters, 164, 171
  implementation, 47
  linear phase, 172
  non-linear phase, 172
Instruction cycle, 124
Integer band sampling technique, 139
Intelligibility, 137–146
Intel 8-bit, microprocessor, 124
Interpolation of stochastic signals, 163–169
Interrogation-reply scheduling technique, 132
Inverse path sequence, 132
Isotropic gradient masks, 263
I.V.R. transformation, 33

K

Kaiser window, 168
Kalman filters, 183
  2-D, 185
Kernel primitives, 242
Kirsch mask, 264

L

Ladder filter, 27
Lagrange interpolator, 168
Lagrangian filter, 163
Laplacian probability density function, 149, 150, 151
Least error energy, 13, 14
Least mean square algorithm, 21–25
Least mean square filter, adaptive, 21–25
Levinson's recursion formula, 14, 18, 38
  algorithm, 35, 37, 44
  factor, 40
Liapunov equation, 186
Limit cycle oscillations, 106, 108, 112
Linear frequency, 229
Linear, interpolation, 166
Linear phase FIR filters, 3–5, 164
Linear phase equalization, 52
Linear transformation, 27–33
Linear transformation matrix, 27, 30, 31
Linearly transformed two port, 28
Listening, informal, 144, 145
Local operators, 261, 262
Low order filters, 3–5
Low pass filter 3, 55
Liu decomposition, 12

M

Mask,
 isotropic, 263
 Kirsch, 263
 Prewitt, 264
 Robinson, 267
 smoothed, 263
 Sobel, 263
Matrix,
 chain, 28
 linear transformation, 27, 30, 31
 Pick, 44
Maximum phase factor, 12
McClellan program, 3, 4, 7, 8
Mealy model, 110
Mean square error, 21–25, 165–168
Microprocessor, digital filter implementation, 123
 Intel 8-bit, 124
Minimum phase factor, 12, 15
Minimum phase filters, 48, 55, 56
Mirror image polynomials, 12, 13
Modulo numbers, 229–237
Monocentric space perspectives, 272
Monodimensional Fourier transform, 208
Monotonic signals, 254
Monotonic variations, 255
Monte Carlo methods, 217–220, 227
 fast, 218
Moving target indicator, 222
 radar, 224
MTA transformations, 33
Multi-dimensional quantization, 117 118
Multi-passband, 4–6
Multi-stopband, 4–6
Multi-rate systems, 139, 142, 146
Mutual inductance, 32

N

Narrow-band conditions, 75, 79
Narrow-band process, 80
Network consideration, speech 138
Noise power spectrum, 212
Non-Gaussian, 77, 81
Non-linear equations, 12
Non-linear optimization techniques, 50
Non-recursive 2-D digital filter, 262
Numbers, Fermat, 66
Number theoretic transforms, 61, 66, 111, 232
Nyquist, frequency, 7, 210
Nyquist rate, 90, 229

O

Ocean bottom reflection coefficients, 87–95
One-dimensional Fourier transform, 87–95
One dimensional space, 118
Optimal filter, 164–167, 178
Optimal linear predictor, 35
Optimization of measurements, 276
Orientation matrix, 276
Orthogonal filters, 35
Overload, 148

P

Packing efficiency, 244
Partial differential equation, 213
Partition, 118
Passband filters, 3, 4
Periodic spectrum, 142
Phase equalization, 52
Phase filter,
 all pass, 48
 2-D, 47–57
 design, 50
Phase response optimization, 50
Pick matrix, 44
Piecewise linear approximation, 150
Pixel images, 209
Planimetric perspective, 271
Plane waves, inhomogeneous, 91, 91
Plane wave reflection coefficient, 87
Poles and zeros, 174–175
Polynomial filters, 8
Polynomials,
 finite, 110
 multivariable, 112
Power variation in seismology, 252–256
Prediction, stochastic, 35
Prewitt mask, 262
Prime factor algorithm, 66
Probability density, 118

Processes,
  external, 242
  internal, 242
Processor, 242
Projection number, 209
Projection-slice theorem, 87, 88, 197
Pseudo-Fermat transformation, 68
Pseudo-Mersenne transform, 68
Pulse amplitude modulation, 148, 149
Pulse code modulation,
  A-Law, 147, 150–154, 161
  coder, 147–161
  μ-law, 147, 148, 150–161

Q

Quantization, 147–161
  in amplitude, 105, 147
  in time, 105
  in loss, 120
  noise, 147
  1-D, 118, 120
  multi-dimensional, 117, 118
  2-D, 117–123
  uniform, 118
  vector, 117, 118
Quine-McCluskey technique, 107

R

Radar systems, 217–227
  constant false alarm rate (CFAR), 218, 221, 222
Radix-2, FFT 65, 67
Radix-3 FFT, 61–68
Radix-4 FFT, 65
Rate-distortion bound, 120
Recursive filters, 22–25
  adaptive LMS, 21
  design, 50
  first order, 106
  2-D, 47–50
  polyphase, 8
Reduced variance methods, 218
Reflection coefficients, of ocean bottom currents, 87–95
  plane wave, 87
Reflected waves, 258
Remez exchange algorithm, 3, 4, 8
Restoration of scintigraphic images, 183–196
RF channel, 240
Riccati equation, 188
Ring of numbers, 61, 65
Ripple, 4, 7
Rivard's algorithm, 97, 101
Robinson mask, 267
Robustness, 142, 144
Roll-call scheduling, 243, 247
Round-off errors, 18

S

Sample density, 2-D, 199–201, 202
Sample density, 3-D, 201–203
Sampling techniques, applications to radar systems, 217–227
Scattering matrix, 36
Scheduling algorithm, 244
Schwarz's Lemma, 42
Scintigraphic images, 183–196
Semaphores, 242
Seismic signal processing, 48, 53, 54
Seismology, 251–258
Seismometer, 251–258
Shifted DFT, 69–73
  direct, 70
  reverse, 70
Signal degradation, 147
Signal flow diagram, 28, 29, 31, 32
Sliding window, 253
Smoothed gradient masks, 263
SNR (signal to noise ratio), 147, 148, 150–154
Sobel gradient masks, 263
Software functions for channel management, 242–243
Space domain, 47
Spatial filtering, 268
Spatial wavenumber, 92
Spectral density, 82
Spectral estimation, 75–85
Spectral factorization, 11–19
Spectral flattening, 139–143
Specular angle, 90–91
Specular reflection, 91
Speech, 148, 150–161
  processing, 137–146
  quality, 137
Squared-error fidelity criterion, 118, 119, 121

Squared-magnitude function, 17
Stability, of finite state machines, 112, 113
of linearly transformed filters, 30
Standard DFT, 70–73
Standard Monte Carlo technique, 217, 218
State-space model, 186, 187, 190
Stationary process, 76
Stochastic, prediction, 35
Stochastic processes, 75
Stochastic signals, 163–169
Stop band filters, 2
Stratified sampling techniques, 217, 218
Strip filtering techniques, 195
Sub-band coder, 139, 142, 144

T

Target list update, 243
Telephone, 137
Termination, 29
Three-dimensional mapping, 233
Three-dimensional reconstruction, 197–205
Three-dimensional representation, 261
Three-dimensional sample density, 199–202
Three-dimensional weighting function, 201
Time division multiplex, 171
Time scheduling procedure, 128, 129
Time weighted flow graph, 125, 127, 131
Tissue absorption, 212–214
Toeplitz forms under narrow band conditions, 75–85
Toeplitz matrix, 12, 14, 187–189
Tomographic system, 207–216
Tomography, 197
Topological implementation, 123–133
Topological limitations, 125
Trapezium projection, 271
Transaction, preparation, 242
update, 243
Transfer time, 127
Transform,
Fourier–Bessel, 88
Hankel, 88, 89
pseudo-Fermat, 68
pseudo-Mersenne, 68
Transformation,
bilinear, 30–32
I.V.R., 33
linear, 27–33
M.T.A., 33
Transmission, high-speed data, 25
Transversal LMS filter, 21, 25
Transversal filter, 229
Truth table, 107–108
Two dimensional convolution, 231
Two-dimensional digital filter, 262
Two-dimensional frequency filtering, 261
Two-dimensional Fourier transform, 87–95
Two-dimensional FFT, 97–102
Two-dimensional Kalman filtering, 183–196
Two-dimensional mapping, 230, 232
Two dimensional phase filters, 47–57
Two-dimensional quantization, 117–121
Two-dimensional recursive filter, 47, 48, 49
Two-dimensional representation, 261
Two-dimensional sample density, 201–203
Two-dimensional signals, 70
Two dimensional weighting function, 201
Two dimensional z-transform, 184
Two-input adder, 128, 131
summing node, 125
Two port, linearly transformed, 28
Two-stage decimated chirp filter, 237

U

UHF radio, 239
Uniform quantization, 118
Up link message, 239, 241–242, 244

V

Variable rate speech processing, 137–146
Variable-transfer time, 126, 128
Vector quantization, 117, 118
Virtual decision levels, 148
Vocoder channel, 137, 139, 142

W

Wave digital filters, 30, 36–38
Wavenumber, 92, 24
Weighting factor, 219, 220
Weighting function, 200
2-D, 201
3-D, 201
Wiener filter, 184
Window coefficients, 82, 83
Window, sliding, 253
Word length, 124

Z

z-domain, 188
z-transform, 35, 105, 188
2-D, 184